Scientific Basis of Healthcare
Angina

Scientific Basis of Healthcare

- AIDS and Pregnancy
- Angina
- Arthritis
- Asthma

Editorial Advisory Board

Prof. Sally Wai-chi Chan (Singapore)
Prof. Caroline J. Hollins-Martin (UK)
Prof. Diana T.F. Lee (Hong Kong)
Dr. Jean Rankin (UK)

Scientific Basis of Healthcare
Angina

Editors

Colin R. Martin PhD
Chair in Mental Health, School of Health
Nursing and Midwifery
University of West of Scotland
UK

Victor R. Preedy PhD DSc
Professor of Nutritional Biochemistry
School of Medicine
King's College London
and
Professor of Clinical Biochemistry
King's College Hospital
UK

 Science Publishers
Jersey, British Isles
Enfield, New Hampshire

Published by Science Publishers, an imprint of Edenbridge Ltd.
- St. Helier, Jersey, British Channel Islands
- P.O. Box 699, Enfield, NH 03748, USA

E-mail: *info@scipub.net* Website: *www.scipub.net*

Marketed and distributed by:

CRC Press
Taylor & Francis Group
an informa business
www.crcpress.com

6000 Broken Sound Parkway, NW
Suite 300, Boca Raton, FL 33487
270 Madison Avenue
New York, NY 10016
2 Park Square, Milton Park
Abingdon, Oxon OX14 4RN, UK

Copyright reserved © 2012
ISBN 978-1-57808-732-7

Cover Illustrations: Reproduced by kind courtesy of the undermentioned:
- Archives of Internal Medicine; The Prevalence of Weekly Angina Among Patients With Chronic Stable Angina in Primary Care Practices, Sept 14, 2009, 169(16): 1491-9. "Copyright © (2009) American Medical Association. All rights reserved." (Figure No. 3 of Chapter 1 by John F. Beltrame)
- Toshio Imanishi (Figure 1 of Chapter 3)

```
         Library of Congress Cataloging-in-Publication Data
Scientific basis of healthcare. Angina/editors, Colin R. Martin,
Victor R. Preedy. -- 1st ed.
       p. ; cm.
  Angina
  Includes bibliographical references and index.
  ISBN 978-1-57808-732-7 (hardcover : alk. paper)
  I. Martin, Colin R., 1964- II. Preedy, Victor R. III. Title:
Angina.
  [DNLM: 1. Angina Pectoris. 2. Evidence-Based Medicine--methods.
3. Myocardial Infarction.  WG 298]
  362.196'122--dc23
                                                        2011038877
```

The views expressed in this book are those of the author(s) and the publisher does not assume responsibility for the authenticity of the findings/conclusions drawn by the author(s). No responsibility is assumed by the publisher for any injury and/or damage to persons or property as a matter of products liability, negligence or otherwise, or from any use or operation of any methods, products, instructions or ideas contained in the material herein. Because of rapid advances in the medical sciences, in particular, independent verification of diagnoses and drug dosages should be made.

All rights reserved. No part of this publication may be reproduced, stored in a retrieval system, or transmitted in any form or by any means, electronic, mechanical, photocopying or otherwise, without the prior permission of the publisher, in writing. The exception to this is when a reasonable part of the text is quoted for purpose of book review, abstracting etc.

This book is sold subject to the condition that it shall not, by way of trade or otherwise be lent, re-sold, hired out, or otherwise circulated without the publisher's prior consent in any form of binding or cover other than that in which it is published and without a similar condition including this condition being imposed on the subsequent purchaser.

Printed in the United States of America

Foreword

Angina, within the context of cardiovascular disease, represents an enduring and complex issue for healthcare provision and clinical management within contemporary society. Often defined within the public consciousness against a background of anxiety, fear and foreboding, the emerging evidence base provides a significant contrast to such perspectives, eliciting a constellation of insights that taken together enhance the treatment and outcomes of those presenting with this diagnosis. The veracity and clarity of recent research observations also promotes a broader public health message in terms of effective health education and promotion and cardiovascular prevention and rehabilitation. A fundamental problem for those working clinically in this area is the availability of relevant, useful and up-to-date information. Co-morbidities associated with angina represent one of the most complex interactions encountered by modern medicine, yet information within this field is often polarised to the discrete dimensions of pathology associated with angina specifically. Colin Martin and Victor Preedy have done an excellent job bringing together experts in the angina field to produce an accessible, evidence-based and clinically relevant account of the germane issues. Readers with either a clinical or research interest in this area will undoubtedly find this volume a valuable and welcome resource. In summary, I would highly recommend this text as an informative and authoritative account of central issues surrounding angina.

Professor David R. Thompson

Professor Thompson co-heads the Psychosocial Research program at the Cardiovascular Research Centre at the Australian Catholic University. Professor Thompson is a leading International authority in cardiovascular disease and has published over 500 papers, as well as 40 book chapters and 15 books.

Preface

In the past three decades there has been a major sea change in the way healthcare is taught and implemented. Teaching in the healthcare professions have been replaced from a "this is what you do" approach to "this is the scientific basis" ethos of evidence-based material. The healthcare professional is now more educated, more informed and more aware that the foundation of good health is good science. Healthcare practitioners with research doctorates or Masters degrees are also more commonplace. As a consequence, practice and procedures are continually changing with a corresponding improvement in healthcare. Concomitantly, the demand for comprehensive and focused evidenced-based text and scientific literature covering single areas of healthcare or treatments have also increased. Hitherto, these have been difficult to obtain and thus it was decided to work on a collection of books on **The Scientific Basis of Healthcare**. The chapters impart holistic information on the scientific basis of health and covers the latest knowledge, trends and treatments. The ability to transcend the intellectual divide is aided by the fact that each chapter has:

- An *Abstract*
- A section called *"Practice and Procedures"* where relevant
- *Key Facts* (areas of focus explained for the lay person)
- *Definitions of words and terms*
- *Summary points*

The books, each on a different medical condition, cover a wide number of areas. The chapters are written by national or international experts and specialists.

In **Angina** we cover a wide range of areas and subtopics including overviews, angina in the context of other diseases, coronary plaque rupture, angina questionnaires, American College of Cardiology/American Heart Association guidelines, nurse specialists role, non-invasive coronary angiography, drugs used in angina, thrombin inhibitors, external counterpulsation therapy and many more scientific fields associated with **Angina.**

The books are designed for practicing health care workers, trained nurses, nursing students, doctors and medical students, therapists, trainees and practitioners of all health- related disciplines including physiotherapists, midwives, dietitians, psychologists, and so on. The special feature of the book also means it is suitable for post graduates, special project students, teachers, lecturers and professors. It is also suitable for college, university, nursing, and medical school libraries as a reference guide.

Colin R. Martin
Victor R. Preedy

Contents

Foreword v
Preface vii

1. Epidemiology of Angina 1
   John F. Beltrame

2. Chest Pain: Cardiac and Non-Cardiac Causes 21
   Geraldine Lee

3. Coronary Plaque Rupture in Patients with Acute Coronary Syndrome 39
   Toshio Imanishi

4. The Use and Application of Angina Questionnaires 58
   Norul Badriah Hassan and *Muhammad Termizi Hassan*

5. American College of Cardiology /American Heart Association Guidelines for ST-Elevation Myocardial Infarction—Overview of Guidelines for Care Before, During, and After STEMI 74
   Rohit S. Loomba and *Rohit R. Arora*

6. Non-Invasive Coronary Angiography with Cardiac CT in Patients with Angina Pectoris 95
   Christof Burgstahler, Harald Brodoefel and *Stephen Schroeder*

7. Drugs Used in Angina: An Overview 114
   Mario Marzilli and *Alda Huqi*

8. Current Clinical Application of Direct Thrombin Inhibitors in Angina Pectoris 133
   Bernardo Cortese and *Marco Centola*

9. Anti-Anginal Drugs in Focus: Trimetazidine 155
   Mario Marzilli and *Alda Huqi*

10. **Enhanced External Counterpulsation Therapy in Coronary Artery Disease Management**   174
  Ozlem Soran and *Debra L. Braverman*

Index   199
About the Editors   201
Color Plate Section   203

Epidemiology of Angina

John F. Beltrame

ABSTRACT

The term 'angina' refers to a strangling sensation; however it also used in a generic context to describe a group of syndromes manifesting as a strangling chest sensation (angina pectoris) that arises as a result of myocardial ischaemia. Although there are a number of angina syndromes, the most prevalent form is chronic stable angina, which is the focus of this chapter. This condition is characterized by angina pectoris that predictably occurs with exertion and promptly resolves with rest.

The prevalence of stable angina in developed countries is approximately 5% of the male and 4% of the female population. The incidence of new cases in these countries is approximately 49/100,000 men and 28/100,000 women. The incidence and prevalence of stable angina varies with age, gender, geographic region and ethnicity.

Assessment of patients with stable angina requires an evaluation of the patient's symptom status (angina), the impact of the condition on quality of life and the risk of cardiac events. Despite a number of therapies available for the treatment of angina, almost 1 in 3 patients with chronic stable angina experience chest pain at least once a week. Women with stable angina and those with co-existing heart failure or peripheral artery disease often experience frequent angina. Furthermore, the frequency of angina may be a clinical marker of the impact of this condition on the

---

The University of Adelaide, Discipline of Medicine, The Queen Elizabeth Hospital Campus, 28 Woodville Rd, Woodville South, Adelaide, South Australia, Australia; Email: john.beltrame@adelaide.edu.au

List of abbreviations after the text.

patient's quality of life. Although stable angina patients do not have as high a risk of cardiac events as those with a recent myocardial infarction, they do have an annual risk of myocardial infarction of <2%/yr and of all-cause mortality of about 2%/yr.

INTRODUCTION—DEFINING ANGINA

The term 'angina' has been used since the 16th century and unfortunately over time it's meaning has become ambiguous. Thus before embarking on a scientific discussion regarding its epidemiology, it is necessary to clarify its definition.

The word 'angina' is derived from the Greek word 'ankhon' which refers to a strangling sensation. Accordingly, it has been used to describe any symptom that is characterized as a strangling sensation such as 'Ludwig angina' where a strangling sensation in the throat is due to a soft tissue infection in the floor of the mouth. Thus the more specific term for the cardiac symptom is 'angina pectoris', where the term 'pectoris' is derived from Latin and refers to the chest.

In 1772, William Heberden published the original description of 'angina pectoris' in his paper 'Some account of a disorder of the breast' (Heberden, 1772), where he described it as:

> "They who are afflicted with it, are seized while they are walking, (more especially if it be up hill, and soon after eating) with a painful and most disagreeable sensation in the breast, which seems as if it would extinguish life, if it were to increase or to continue; but the moment they stand still, all this uneasiness vanishes."

This description is as valid today as when it was first penned and aptly describes what is referred to as 'exertional angina'. However angina pectoris may also occur without physical activity whereupon it is referred to as 'rest angina' (Maseri 1995). Some patients may describe the pain occurring during exertion on some occasions but at rest on others. In these situations, it is referred to 'mixed-pattern angina' (Maseri 1995). Alarmingly, some patients describe a building-up in the frequency of angina referred to as 'crescendo angina' where the pain occurs more often and/or with less exertion.

MYOCARDIAL ISCHAEMIA AND ANGINA SYNDROMES

The ambiguity of the term 'angina pectoris' has arisen from its use to describe a group of clinical disorders rather than a symptom. In contemporary medicine, 'angina' is often used to refer to disorders that cause myocardial ischaemia and manifest as chest pain. This broader conceptual definition of 'angina' as chest pain attributable to myocardial ischaemia requires further clarification.

Myocardial ischaemia is a pathological condition where an insufficient blood supply results in inadequate oxygen supply and the accumulation of wastes products in the myocardium. This can arise when there is an increased oxygen demand by the myocardium (e.g., during exertion) and/or impaired blood supply by the coronary circulation. Thus myocardial ischaemia typically arises from dysfunction of the small or large coronary blood vessels; generically referred to as coronary heart disease (CHD). The later term not only includes reversible myocardial ischaemia but also myocardial infarction where the prolonged ischaemic insult has resulted in irreversible myocardial cell death.

Clinically, several anginal syndromes have been described and these are summarized in Table 1.1. *Chronic stable angina* is due to obstructive coronary artery stenoses, which predictably produce myocardial ischaemia when excessive myocardial oxygen demand occurs and thus manifest as exertional angina.

In contrast, *unstable angina* occurs when a coronary artery suddenly occludes (either partially or completely), typically due to a coronary thrombus, resulting in an abrupt reduction in blood flow. Myocardial ischaemia rapidly ensues and may quickly progress on to myocardial infarction unless blood flow is restored. This form of angina typically manifests as rest angina although in its early phases may also be present as crescendo or mixed pattern angina.

In 1959, Prinzmetal (Prinzmetal et al. 1959) described another form of angina that manifests as recurrent episodes of rest angina associated with ST elevation, which promptly responded to nitrates. The mechanism responsible for this anginal syndrome was coronary artery spasm and is referred to as *variant angina*.

Table 1.1 Types of Angina.

Angina Syndrome	Clinical Features
Unstable Angina	• Characterized by crescendo or rest angina • An acute coronary syndrome manifestation with high risk of progressing to myocardial infarction • Typically due to an unstable atherosclerotic plaque
Stable Angina	• Characterized by exertional angina • Typically due to a tight obstructive coronary artery stenosis
Prinzmetal Variant Angina	• Characterized by rest or nocturnal angina • Typically due to coronary artery spasm
Decubitus Angina	• Characterized by angina when lying down • Typically due to left ventricular dysfunction resulting in redistribution of pulmonary fluids and thus increased cardiac workload
Silent Ischaemia	• Absence of angina in the presence of documented ischaemia • May occur with coronary artery or microvascular dysfunction
Syndrome X	• Includes classical syndrome X, microvascular angina, and the coronary slow flow phenomenon. • Characterized by prolonged episodes of exertional or rest angina • Typically due to coronary microvascular dysfunction

In 1973, Kemp (Kemp et al. 1973) coined the term '*syndrome X*' to describe a new anginal condition characterized by exertional angina, a positive exercise stress test for ischaemia yet normal coronary angiography. This form of angina is attributed to coronary microvascular dysfunction and remains a diagnostic and therapeutic enigma.

Decubitus angina is characterized by chest pain (often with dyspnoea) occurring when the patient lies down. This condition requires considerable further research as its mechanism is not fully elucidated but believed to be due to left ventricular dysfunction, especially diastolic dysfunction.

Myocardial ischaemia may also occur in the absence of angina and is thus referred to as *silent ischaemia*. Although not truly an 'anginal syndrome', it is frequently described in this context as it reflects a defective anginal warning system.

AMERICAN COLLEGE OF PHYSICIANS DEFINITION OF ANGINA

Despite the diverse implications of the term 'angina' (as described above), when most clinicians use the term angina, they are referring to patients with chronic stable angina that manifests as the classical exertional angina as initially described by Heberden. Indeed an operational definition of what represents angina has been proposed by the American College of Physicians (Diamond 1983) and is summarized in Table 1.2. This definition describes angina as either 'typical' or 'atypical' on the basis of how many of the clinical features are consistent with exertional angina. In those patients with features of typical angina, the sensitivity and specificity for detecting significant coronary artery disease on angiography is respectively 91 and 87% in males, and correspondingly 89 and 63% in females (Detry et al. 1977).

Table 1.2 American College of Physicians Angina Pectoris Definition.

Chest Pain Features
1. Substernal chest discomfort—characteristic quality (tightness) & duration (minutes)
2. Provoking Factors—exertion or emotional stress
3. Relieving Factors—rest or sublingual nitrates
Classification
Typical Angina—all 3 of above criteria met.   **Atypical Angina**—only 2 of above criteria   **Non-cardiac Chest Pain**—only 1 of above criteria

In this chapter, the data presented concerning 'angina' will primarily reflect patients with chronic stable angina. It will particularly focus on the prevalence, incidence, clinical profile, associated morbidity and mortality with this condition. Although the clinical features of the other forms of angina have been alluded to, their epidemiological aspects are less clearly described and unfortunately there are no studies that directly compare the prevalence or incidence of the various forms of angina.

THE PREVALENCE AND INCIDENCE OF ANGINA

The prevalence of a condition refers to its frequency within a given population at a particular point in time. In 2009 in the United Kingdom, it was estimated that 2.1 million people suffered from angina thus representing a prevalence of approximately 5% of men and 4% of women (British Heart Foundation 2010). In the United States, approximately 10.2 million Americans were reported to have angina in 2006 with 4.7% of Caucasian men and 4.5% of Caucasian women over the age of 20 yr affected (American Heart Association 2010). These data are primarily based upon patient self-report of a history of angina and thus subject to limited validity.

Although the prevalence of angina in the above populations is similar, it is affected by age, gender, ethnicity, and geographic region. As shown in Fig. 1.1, within the United Kingdom the prevalence is almost 17% amongst males and 12% in females over the age of 75 yr but is less than 1% of all those under 45 yr of age. Furthermore

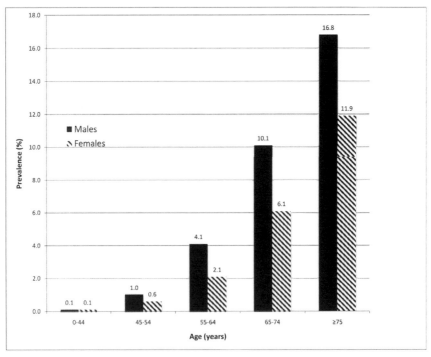

Figure 1.1 Age-specific Prevalence of Angina in the United Kingdom in 2009. (Data derived from British Heart Foundation 2010).

for all ages, the prevalence of angina in men from northern Ireland is approximately 6% whereas amongst Welshman it is 4% (British Heart Foundation 2010). Ethnic differences in angina occurrence is well illustrated in the United States where the prevalence in men over the age of 20 yr is 3.8% in Caucasians, 3.3% in African Americans, and 3.6% in the Hispanic population. The equivalent prevalence amongst females are 3.7, 5.6 and 3.7% (Roger et al. 2011).

The incidence of a condition refers to the number of new cases within a given population over a specified period of time. Thus although approximately 2.1 million Britons suffered from angina (prevalence) in 2009, there is only estimated to be 28,000 new cases (British Heart Foundation 2010). These estimates are calculated by sampling a group of general practitioners and reviewing their patient case records. Thus considering all ages, approximately 49/100,000 males were newly diagnosed angina cases and 28/100,000 females.

CLINICAL CHARACTERISTICS OF STABLE ANGINA PATIENTS

Unlike acute coronary syndromes where there are multiple registries of patients with this well defined condition, there are few large registries of patients with stable angina given the difficulty in defining the condition. The largest representative cohort of stable angina patients are the 2,031 patients recruited to the CADENCE (Coronary Artery Disease in gENeral practice) (Beltrame et al. 2009). In this study, participating Australian general practitioners were asked to record clinical details on 10–15 consecutive 'angina patients' seen during their routine consultations, irrespective of the reason for the consultation. The study cohort was shown to be representative of the Australian population and general practitioners recruitment of these 'angina patients' appropriate since almost three-quarters had American College of Physicians (ACP) criteria for angina, 93% had objective evidence of coronary artery disease, and 89% had been seen by a cardiologist (Beltrame et al. 2009). The objective of the study was to profile health status and management of a representative stable angina cohort and thus patients were also asked to complete a health-related quality of life questionnaire.

The CADENCE study provides some insights into the clinical characteristics of patients with stable angina and these are summarized in Table 1.3. This cohort of stable angina patients was elderly (mean age = 71 ± 11 year) and two-thirds male as would be expected from the prevalence presented above. Furthermore the mean period from the diagnosis of angina was more than 8.2 ± 7.9 year with over half of the patients having at least a 5-year history. Coronary risk factors were prevalent in these patients with most having a history of cigarette smoking, hypertension, hypercholesterolaemia or obesity, with about one-third having a history of diabetes (Table 1.3). Over two-thirds of patients had previously experienced an episode of unstable angina or a

Table 1.3 Clinical Characteristics in Stable Angina Patients.

Clinical Characteristic	Prevalence
Coronary Artery Disease Risk Factors	
Age	71 ± 11 year
Male gender	64%
Diabetes mellitus	30%
Hypertension	72%
Hypercholesterolaemia	78%
Previous or current smoker	59%
Obesity (BMI and/or waist circumference)	85%
Associated Cardiovascular Disease	
Previous acute coronary syndrome	70%
Cardiac Failure	22%
Peripheral Artery Disease	17%
Atrial Fibrillation	10%
Angina Characteristics	
Substernal chest discomfort	93%
Pain provoked by exertion	73%
Pain provoked by emotional stress	26%
Pain relieved by rest	54%
Pain relieved by sublingual nitrates	51%
American College of Physicians Angina Criteria	72%
CAD Investigations	
Exercise Test	63%
Coronary Angiography	78%
Myocardial Scintigraphy	28%
Stress Echocardiography	26%

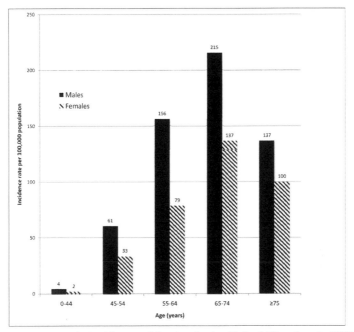

Figure 1.2 Age-specific Incidence of Angina in the United Kingdom in 2009. (Data derived from British Heart Foundation 2010).

myocardial infarct but less than a quarter had co-existing cardiac failure and only 10% had atrial fibrillation. Although not all patients had American College of Physicians criteria for angina, most had reported episodes of chest pain with almost three-quarters reporting exertional pain but only 26% emotion-related chest pain (Table 1.3). About half reported that the pain improved with rest and/or sublingual nitrates. Coronary angiography was the most frequently utilized diagnostic test in the stable angina patients with two-thirds undertaking exercise stress testing and fewer undergoing other functional tests (Table 1.3). Hence this study profiles the typical stable angina patient.

HEALTH OUTCOMES IN STABLE ANGINA PATIENTS

A disease process may affect the patient's health status in a variety of ways. Firstly, the disease manifests as a symptom, which in the case of myocardial ischaemia is angina. This symptom may impinge on the patient's physical ability and hence the angina may limit a patient's physical capacity. Furthermore, the symptoms will affect the patient's quality of life but this will be dependent upon the patient's expectations of their health. For example, angina may frustrate one patient because of the exercise limitations (e.g., not being able to walk to the local shops) as a result of this disease whereas another individual may not perceive this as a problem. Finally the disease process may result in a cardiac event such an acute myocardial infarct or cardiac mortality, both of which may occur in stable angina patients. The following sections will describe the available data concerning the health status of stable angina patients.

ANGINA SYMPTOMS AND HEALTH RELATED QUALITY OF LIFE

The principal finding of the CADENCE study was that 29% of stable angina patients experience an episode of angina at least once a week; a finding consistent with other angina studies (Spertus et al. 2002, Wiest et al. 2004, Kirwan et al. 2008, Kirwan et al. 2005). The independent clinical predictors of patients with at least weekly angina included female gender, heart failure and a history of peripheral vascular disease (Beltrame et al. 2009). The predilection for women having more frequent angina is likely to be multi-factorial and may include biological, clinical presentation and assessment differences between genders (Bairey-Merz et al. 2006). For example, women may have smaller coronary arteries that are less amenable to revascularisation therapies. Furthermore, coronary microvascular dysfunction is more prevalent in women and angina resulting from this is less responsive to conventional anti-anginals. Co-existing heart failure in patients with angina may reflect more extensive coronary artery disease as may peripheral artery disease, hence patients with these disorders have more frequent angina.

The patients who experience angina at least once a week, are more physically limited by their angina and have a poorer quality of life than those patients who experience angina less often (Beltrame et al. 2009). Thus simply enquiring how often a patient experiences angina may provide clinical insights into the impact of this condition on the patient's quality of life. Although the CADENCE study utilized a threshold of angina at least once week, the relationship is a continuum as shown in Fig. 1.3A. Thus the more frequent the angina, the greater the impairment in physical limitation and quality of life.

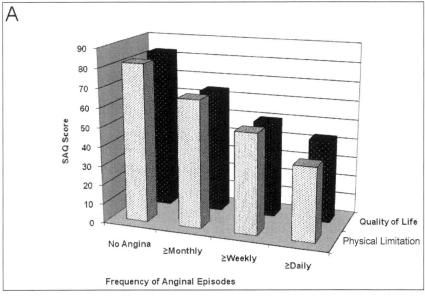
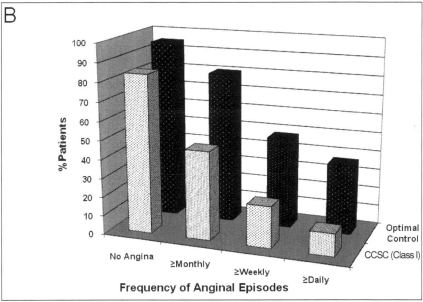

Figure 1.3 Relationship between Angina Frequency and (A) Patient-assessed Quality of Life Indices, and (B) General Practitioner's Perception. The Patient-assessed measures were evaluated by the Seattle Angina Questionnaire. The General Practitioners reported whether they believed the patient's angina was 'optimally controlled' and also classified their physical disability using the Canadian Cardiovascular Society Classification (CCSC). Data derived from Beltrame et al. 2009.

As shown in Fig. 1.3B, this relationship also exists for clinician derived parameters such as the Canadian Cardiovascular Society Classification (CCSC) (Campeau 1976). In the CADENCE study, the clinicians were asked to evaluate if their patient's angina was (a) optimally controlled, and (b) the extent of physical limitation as assessed by the CCSC. Thus as summarized in Fig. 1.3B, patients with less frequent angina were more likely to be considered to have optimally controlled angina and CCSC Class I angina (no physical limitation in activities of daily living by the angina) by their clinicians.

While enquiring about the frequency of angina provides some insights into the disability associated with the disorder, it does not replace a detailed history and evaluation identifying the full impact of the condition on the patient. Unfortunately clinicians may not be completely aware of the angina burden experienced by their patients as alluded to in the CADENCE study. In this study, the clinicians reported that 80% of their patients had optimally controlled angina and that 61% had minimal impairment in their physical activity by the angina. In contrast, from the patient questionnaires, only 52% of patients reported being angina-free and only 47% described their angina as not limiting their enjoyment in life. Hence further efforts are required to bridge this gap between the patient's experience and the clinician's perception of the disability associated with angina.

CARDIAC EVENTS

Patients with chronic stable angina may experience cardiac events including myocardial infarction and death although examining the relationship of these events to stable angina is often difficult. Patients with established angina may subsequently experience a myocardial infarct or CHD death. Alternatively, myocardial infarction may represent the initial CHD presentation with patients reporting angina thereafter. However not all patients who have experienced a myocardial infarct or CHD death necessarily have a history of stable exertional angina.

Only a few follow-up studies have focussed on the frequency of cardiac events in stable angina patients. The Framingham study (Kannel and Feinleib 1972) and Health Insurance Plan study (Frank et al. 1973) were conducted over 30 yr ago and reported an annual myocardial infarct rate of 4–5%/yr and all-cause mortality of 3–4%/yr in angina patients who had not previously experienced a myocardial infarct. A more recent study of middle-aged British men with angina but without a prior history of myocardial infarction, reported 1.7%/yr myocardial infarct rate and 1.9%/yr all-cause mortality (Lampe et al. 2000). Similarly, the Gothenberg study reported 1.8%/year infarct rate and 2%/yr all-cause mortality in Swedish men (Hagman et al. 1988). The lower incidence of cardiac events amongst the angina patients in the more recent studies is possibly attributed to the improved medical care of these patients.

There is extensive data on the incidence and prevalence of cardiac events in large population studies and although it is unclear how many of these patients have stable angina, they do provide some insights into determinants of these events. Approximately 1.5 million Britons reported that they had experienced a myocardial infarct at some point in their life, representing an approximate prevalence of 4.8% amongst men and 2.0% women (British Heart Foundation 2010). Similarly, the prevalence of myocardial infarction in the United States was 4.3% in males and 2.2% in females (Roger et al. 2011). Moreover (as in stable angina), the prevalence of myocardial infarction varies not only with age and gender but also geographic location and ethnicity. For example within Britain, the prevalence on myocardial infarction amongst men in Scotland is 4% whereas in northern Ireland it is 5% and in Wales 6% (British Heart Foundation 2010). Ethnic differences in myocardial infarction have been reported within the United States with only 3.0% of Hispanics reported to have experienced a myocardial infarct as compared with 4.3% of Caucasians or African Americans (Roger et al. 2011).

Coronary heart disease mortality statistics are regularly collected by the World Health Organisation and provide useful insights into the prevalence of the disease across the globe. As shown in Fig. 1.4A, there is considerable variability in CHD mortality between countries varying from 40-50 CHD deaths/100,000 population in the Far eastern countries such as China, Japan and Korea, to over 200 CHD deaths/100,000 population in select eastern European countries, with the Russian Federation recording over 700 CHD deaths/100,000 population. The trends are similar amongst females but with an overall lower prevalence of CHD death (Fig. 1.4B). The reason for this variability is multi-factorial including regional differences data collection and biological differences arising from

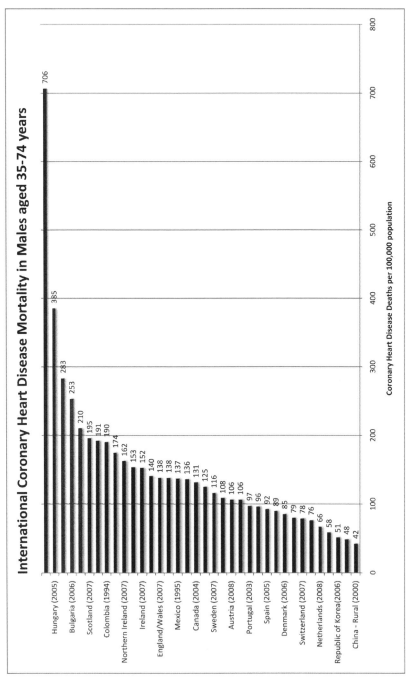

Figure 1.4A Age-standardized Coronary Heart Disease Mortality in Males aged 35–74 yr.

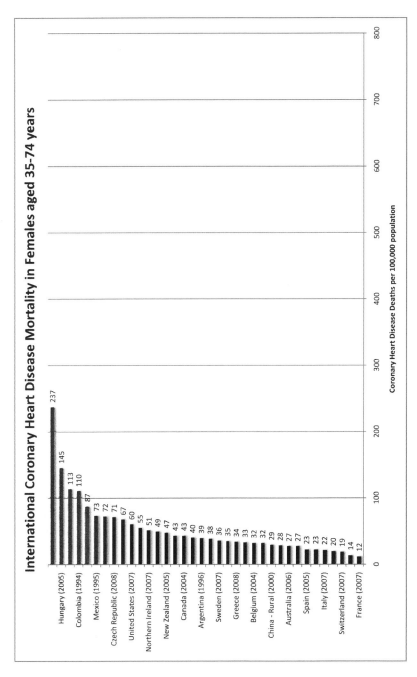

Figure 1.4B Age-standardized Coronary Heart Disease Mortality in Females aged 35–74 yr.

genetic and environmental influences. Unfortunately the same detail is not available for regional prevalence in stable angina, which is more difficult to document.

Coronary heart disease mortality has improved over the past 30–40 yr, showing peak prevalence in many countries in the late 1960's. Figure 1.5 summarizes the CHD mortality within the Australian population over the past 20 yr. As shown in this figure, the improvement is evident in both genders and particularly amongst males. These improvements are believed to be due to community awareness and uptake of important lifestyle changes as well as improved medical therapies. Whether this temporal improvement in CHD mortality is also evident in chronic stable angina is unclear considering the difficulty in defining and monitoring the latter.

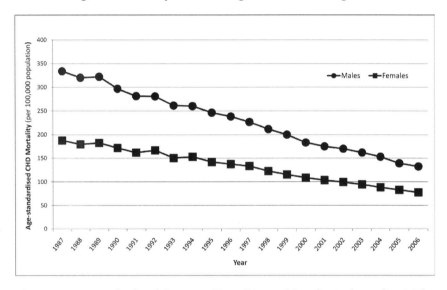

Figure 1.5 Age-standardized Coronary Heart Disease Mortality in Australian Males and Females from 1987 to 2006.*

*Age-standardized mortality rates calculated as age-specific rates applied to a standard population age structure as of 30th June 2001.
#Deaths coded using International Classification of Diseases (ICD)-9 for 1987–1996 and ICD-10 for 1997–2006.

PRACTICES AND PROCEDURES

- The evaluation of angina pectoris requires a detailed history to characterize the type of angina the patient is experiencing and thus determine the associated anginal syndrome.
- Angina frequency is important to evaluate in patients with chronic stable angina as it may reflect the extent of disability experienced by the patient and thus its impact on quality of life.
- Amongst patients with chronic stable angina, women and those with a history of heart failure or peripheral artery disease, are susceptible to experiencing frequent angina.

KEY FACTS

- Chronic stable angina is a condition due to blockages in the coronary arteries that manifests as chest pain during exercise that promptly resolves with rest.
- Approximately 5% of men and 4% of women in developed nations suffer from chronic stable angina.
- In Britain, there are 49 per 100,000 men and 28 per 100,000 women, newly diagnozed with chronic stable angina.
- Patients with chronic stable angina frequently have coronary risk factors including obesity, high blood pressure, high cholesterol levels, smoking and diabetes.
- About 1 in 3 patients with chronic stable angina experience chest pain once a week despite current treatments.
- Women are particularly susceptible to frequent episodes of chest pain.
- In patients with chronic stable angina, the annual risk of experiencing a heart attack is 1–2% and the risk of dying is about 2%.

SUMMARY

- Angina pectoris is a strangling sensation in the chest arising from myocardial ischaemia.

- There are several angina syndromes with the most prevalent being chronic stable angina.
- Chronic stable angina manifests as exertional angina that resolves promptly with rest.
- The prevalence of stable angina in developed countries is 5% of men and 4% of women.
- The incidence of new cases in these countries is 49/100,000 men and 28/100,000 women.
- Almost a third of stable angina patients experience angina at least once a week despite therapy.
- Female stable angina patients and those with heart failure or peripheral artery disease, often have frequent angina.
- The frequency of angina episodes may reflect the impact of this disorder on quality of life.
- Stable angina patients have <2%/yr risk of myocardial infarction and 2%/yr all-cause mortality.

DEFINITIONS

Angina Pectoris: a strangling sensation in the chest resulting from myocardial ischaemia.

Coronary Heart Disease: a group of clinical disorders involving coronary circulatory dysfunction resulting in impaired coronary blood flow and thus myocardial ischaemia. This includes coronary atherosclerosis, coronary artery spasm and/or microvascular dysfunction.

Crescendo Angina: angina pectoris that is occurring more frequently or with greater intensity, or with less provocation. It typically heralds the development of unstable angina.

Exertional Angina: angina pectoris precipitated by exertion.

Incidence: the number of new episodes of a disorder over a period of time (e.g., new myocardial infarct events in 2007).

Mixed Pattern Angina: angina pectoris occurring during exertion but also on occasions at rest.

Myocardinal Infarct: a pathological condition where inadequate coronary blood flow results in myocardial necrosis.

Myocardial Ischaemia: a pathological condition where an insufficient coronary blood flow results in inadequate oxygen supply and the accumulation of wastes products in the myocardium.

Prevalence: the number of patients with the disorder at any particular time (e.g., patients with a myocardial infarct in Britain).

Rest Angina: angina pectoris occurring at rest.

LIST OF ABBREVIATIONS

CADENCE	:	Coronary Artery Disease in General Practice
CCSC	:	Canadian Cardiovascular Society Classification
CHD	:	Coronary Heart Disease
PCI	:	Percutaneous Coronary Intervention

REFERENCES CITED

American Heart Association. 2010. Heart Attack and Angina Statistics. In Heart Disease and stroke statistics. 2010.

Bairey-Merz, C.N., L.J. Shaw, S.E. Reis, et al. 2006. Insights from the NHLBI-Sponsored Women's Ischemia Syndrome Evaluation (WISE) Study: Part II: gender differences in presentation, diagnosis, and outcome with regard to gender-based pathophysiology of atherosclerosis and macrovascular and microvascular coronary disease. J Am Coll Cardiol 47: S21–9.

Beltrame, J.F., A.J. Weekes, C. Morgan, R. Tavella and J.A. Spertus. 2009. The prevalence of weekly angina among patients with chronic stable angina in primary care practices: The Coronary Artery Disease in General Practice (CADENCE) Study. Arch Intern Med 169: 1491–9.

British Heart Foundation Statistics Database. 2010. Coronary heart disease statistics 2010. British Heart Foundation Health Promotion Research Group. Oxford, UK.

Campeau, L. 1976. Grading of angina pectoris. Circulation 54: 522–3.

Detry, J., B. Kapita, J. Cosyns, B. Sottiaux, L. Brasseur and M. Rousseau. 1977. Diagnostic value of history & maximal exercise ecg in men & women suspected of coronary heart disease Circulation 56: 756–61.

Diamond, G.A. 1983. A clinically relevant classification of chest discomfort. J Am Coll Cardiol 1: 574–5.

Frank, C.W., E. Weinblatt and S. Shapiro. 1973. Angina pectoris in men. Prognostic significance of selected medical factors. Circulation 47: 509–17.

Hagman, M., L. Wilhelmsen, K. Pennert and H. Wedel. 1988. Factors of importance for prognosis in men with angina pectoris derived from a random population sample. The Multifactor Primary Prevention Trial, Gothenburg, Sweden. Am J Cardiol 61: 530–5.

Heberden, W. 1772. Some account of a disorder of the breast. Medical Transactions 2: 59–67.

Kannel, W.B. and M. Feinleib. 1972. Natural history of angina pectoris in the Framingham study. Prognosis and survival. Am J Cardiol 29: 154–63.

Kemp, H.G., P.S. Vokonas, P.F. Cohn and R. Gorlin. 1973. The anginal syndrome associated with normal coronary arteriograms. Am J Med. 54: 735–42.

Kirwan, B.A., J. Lubsen and P.A. Poole-Wilson. 2005. Treatment of angina pectoris: associations with symptom severity. Int J Cardiol 98: 299–306.

Kirwan, B.A., J. Lubsen, S. de Brouwer, F.J. van Dalen, S.J. Pocock, R. Clayton, N. Danchin and P.A. Poole-Wilson. 2008. Quality management of a large randomized double-blind multi-centre trial: the ACTION experience. Contemp Clin Trials 29: 259–69.

Lampe, F.C., P.H. Whincup, S.G. Wannamethee, A.G. Shaper, M. Walker and S. Ebrahim. 2000. The natural history of prevalent ischaemic heart disease in middle-aged men. Eur Heart J 21: 1052–62.

Maseri, A. Transient myocardial ischemia and angina pectoris: classification and diagnostic assessment. pp. 451–76. *In:* Ischemic Heart Disease A Rational Basis for Clinical Practise and Clinical Research. 1995. Churchill Livingstone. New York, USA.

Prinzmetal, M., R. Kennamer, R. Merliss, T. Wada and N. Bor. 1959. A variant form of angina pectoris: Preliminary report. Am J Med 27: 375–88.

Roger, V.L., A.S. Go, D.M. Lloyd-Jones, et al. 2011. Heart Disease and Stroke Statistics--2011 Update: A Report From the American Heart Association. Circulation. 123: (in press).

Spertus, J.A., P. Jones, M. McDonell, V. Fan and S.D. Fihn. 2002. Health status predicts long-term outcome in outpatients with coronary disease. Circulation 106: 43–9.

Wiest, F.C., C.L. Bryson, M. Burman, M.B. McDonell, J.G. Henikoff and S.D. Fihn. 2004. Suboptimal pharmacotherapeutic management of chronic stable angina in the primary care setting. Am J Med 117: 234–41.

Chest Pain: Cardiac and Non-Cardiac Causes

Geraldine Lee

ABSTRACT

Angina pectoris is an early manifestation of coronary artery disease and requires urgent investigation to determine the cause. In individuals who experience chest pain and where the tests results for CAD are negative, a diagnosis of 'non-cardiac chest pain' is given. The outcome for these patients include being discharged home from hospital with no further planned follow-up, while other patients may be referred to specialties such as gastroenterology, respiratory and psychiatry for further diagnostic tests. This chapter will review the literature relating to the causes of non-cardiac chest pain and examine how the lack of a definitive diagnosis or a non-cardiac chest pain diagnosis affects the individual and can lead to significant healthcare utilization. The role of rapid access clinics and a multi-disciplinary approach to managing non-cardiac chest pain will be described. The presence of cardiovascular risk factors such as cigarette smoking, a diet high in saturated fats and low levels of physical activity should also be examined as many of these modifiable behavioural activities contribute to CAD and other co-morbidities. Recommendations that can be applied into clinical practice will be presented with an emphasis on reassuring patients, undertaking anxiety and depression screening and performing a cardiovascular risk assessment.

Preventative Health, Baker Heart & Diabetes Research Institute, St Kilda Road Central, Melbourne, VIC 8008, Australia; Email: Geraldine.Lee@bakeridi.edu.au

List of abbreviations after the text.

INTRODUCTION

Angina pectoris is a symptom and the most common early manifestation of Coronary Artery Disease (CAD) (Russell et al. 2010). Angina is characterized by central chest pain that may or may not radiate to the left arm, neck and jaw. If a person is asked to score their pain (on a scale of 1 to 10 with 10 being the most severe), the majority would score 8 or 9. Sublingual nitroglycerin may provide some relief for chest pain that is cardiac in origin (although this is not always the case and in some cases for those with gastro-intestinal complaints may also report relief). Once angina has been formally diagnosed as CAD related, the primary objective is to instigate urgent treatment to preserve the affected myocardium and prevent myocardial infarction (MI), heart failure and death.

Often, people will use the term 'angina'/'angina pectoris' or chest pain and we know now that are as many non-cardiac types of chest pain which could explain the symptoms. It may be that both the public and healthcare professionals use the term 'angina' prior to confirmation of the diagnosis. Thus, in this chapter the term 'chest pain' will be used unless specifically referring to angina in the context of CAD.

DETERMINING THE CAUSE OF CHEST PAIN

The primary objective with any person who experiences chest pain is to investigate and treat the cause in order to alleviate the symptoms. This usually involves taking a detailed history with special attention to the onset of chest pain and if anything precipitated the pain, a description and location of the pain, what resolves it and exacerbates it, what the individual was doing when the chest pain started, any previous episodes, other symptoms and their medical history with a focus on cardiovascular (CV) risk factors such as hypertension, hypercholesterolaemia, cigarette smoking, diet (high in saturated fats) and being overweight or obese, low levels of physical activity and the presence of diabetes mellitus. All of this information is important as part of the medical history as well as the age, gender and family history of cardiovascular disease (CVD) including MIs and stroke.

PRACTICE AND PROCEDURES

Documenting the symptoms and the nature of the pain is very important as it may lend some vital clues to the aetiology. Healthcare professionals are educated to associate the terms 'vice-like grip', 'someone standing on my chest', 'red hot pain' with ACS and MI. However, there are many atypical presentations of chest pain which may lead the healthcare professional to dismiss or reconsider a cardiac diagnosis. This is especially relevant when a person experiences pain isolated to their right arm or those who complain of a tight feeling in their throat or flu-like symptoms. Research suggests that women in particular, do not experience typical angina but are more likely to present with atypical symptoms (Ghezeljeh et al. 2010). The lack of homogeneity in how patients describe the location and intensity of their chest pain is an important consideration for healthcare professionals.

As outlined in the table below, there are some important questions to ask a patient.

Table 2.1 Questions to ask patients during clinical assessment.

TOPIC	Questions
LOCATION	Where is your chest pain? Get the patient to show where the pain is
CHARACTER OF PAIN	Describe the pain*
ONSET and DURATION	What were you doing when the pain started?
ASSOCIATED FEATURES	What makes the pain worse and are there any other symptoms?
RELIEF	What relieves the pain?
HISTORY	Have you had this pain before?

*By asking the patient to describe the pain in their own words, it is less likely that assumptions will be made or that the patient will answer yes to any specific questions about the pain that may actually be contradictory.

The clinical examination is an integral part of assessment and should include all the aspects outlined in Table 2.2 below. Once the clinical assessment, medical history and diagnostic tests are carried out and results obtained, a working diagnosis can be made. There are several algorithms available in guidelines for testing those with chest pain including the NICE clinical guidelines (NICE 2010). The key point about this process is that the aim is to establish the *likelihood* of CAD. However all risk prediction rules have limitations and cannot be the sole source on which to base a diagnosis (Gerber et al. 2010).

Table 2.2 Clinical and diagnostic tests to investigate chest pain.

Clinical assessment checklist:
√ Blood pressure
√ 12-lead electrocardiogram
√ Assessment of heart and lungs sounds
√ oedema (peripheral and pulmonary)
√ Chest x-ray
√ Cardiac biomarkers (Troponin I)

Therefore risk prediction scores should be used in conjunction with medical history, clinical assessment findings and biomarker results. For example, for individuals with a positive family history, several cardiovascular risk factors and a 12 lead ECG demonstrating ischemia, their likelihood of CAD is quite high and thus further tests such as coronary angiography are indicated. A recent paper investigated the value of the Framingham and SCORE risk models, risk factors and biomarkers (B-type natriuretic peptide and high sensitivity C-reactive protein) in those with CAD who were referred for elective angiography and age, male gender, diabetes, chest pain and prior CVD were identified as independent predictors of CAD (Kotecha et al. 2010). The predictive ability of these variables is not surprising given that age, male gender and diabetes are known CVD risk factors. Of note is the identification of chest pain as an independent predictor and it also reconfirms chest pain as an important symptom of CAD.

The treatment of angina that is CAD related is well established and the chapters on the American Heart Association guidelines on treating unstable angina and drugs used for angina will provide greater detail. As well as being a manifestation of CAD, angina can also be a symptom of valvular heart disease, cardiomyopathy, pericarditis and aortic dissection. Persistent chest pain can also be due to microvascular coronary disease and abnormal cardiac nociception (Phan et al. 2009). Where a cardiac aetiology is suspected, patients will require echocardiography and in some instances computerized tomography (CT) can be used to determine the structural and functional integrity of the heart and its structures.

CHEST PAIN THAT IS NOT ANGINA

For those who experience chest pain and with the majority of the public aware that chest pain is an ominous sign of a heart attack, many will be present at an Emergency Department (ED) for assessment. Increases in angina and chest pain presentations to the ED over a 10 yr period (from 1990 to 2000) have been reported. One study reported that angina rates had increased by 79% and chest pain by 110% (Rosengren 2008). These increases support the need to examine presentations and treat confirmed CAD-related cases appropriately. There is now a body of evidence demonstrating that many chest pain presentations that were thought to be angina are in fact non-cardiac chest pain. However, a full assessment and working diagnosis is essential and for those who are very unlikely to have CAD (i.e., no risk factors, negative Troponin result and normal ECG), a differential diagnosis should be considered with further assessment for other possible pathologies. The main alternative conditions to consider are pulmonary, gastrointestinal, musculoskeletal, neurological and psychogenic and it is necessary to undertake further assessments so that patients can be given a diagnosis which will hopefully reduce the need for further investigations and reassure the individual (Mayou et al. 1994). Healthcare professionals need to be aware of the how to manage the patient with non-cardiac chest pain as often their ED presentation is seen as episodic and in isolation with no strategic management plan or further referrals to other specialties undertaken.

The primary aim of the ED is to identify and provide rapid treatment/referral for the presenting symptoms rather than identify CAD. The lack of a diagnosis may result in anxiety about their chest pain, no confirmed diagnosis and uncertainty about what they need to do in terms of further episodes and being uncertain if they are at risk of a heart attack. Other pathologies need to be considered.

Pulmonary Conditions

Pulmonary embolism can cause chest pain and dyspnoea and investigations for possible deep vein thrombosis and a previous history of clotting disorder/abnormalities should be performed. Other differential diagnoses to be considered are: pleurisy, pneumothorax and more rarely pulmonary infarction. With suspected pneumothorax, the onset of pain is usually sudden and an absence of breath sounds and hyper-resonance on the affected side are also observed.

Gastro-intestinal Conditions

GI conditions are prevalent in the population and diagnoses including oesophageal spasm or oesophageal reflux need to be considered. Reflux is often worse after eating and described as burning pain and also when the individual lies down, tests to measure gastric acidity and motility should be undertaken. Introduction of antacids and proton pump inhibitors may help relieve symptoms and patients will need reassurance that their symptoms are gastrointestinal in nature with no cardiac component. Confusingly, nitrates may relieve the pain of oesophageal spasm and therefore further investigations are required.

Musculoskeletal Conditions

Another possible cause of non-cardiac chest pain is musculoskeletal. An inflammatory condition known as costochondritis appears with a localized pain over the costochondral junction and worsens with movement and deep inspiration. After exertion, the pain can worsen lasting for several hours. It can be caused by recent viral illness. Other musculoskeletal conditions to consider include cervical spondylosis and degenerative spinal conditions which may produce pain that radiates down the arm. One important assessment finding in musculoskeletal conditions is associated tenderness over the affected area.

Neurological Conditions

Although very uncommon, herpes zoster can produce a band-like chest pain across the upper chest. This pain usually appears in the prevesicular phase of shingles.

Psychogenic Conditions

This differential cause of non-cardiac chest pain is probably the least understood and the most challenging for healthcare professionals. Amongst the possible disorders to be considered are anxiety, hyperventilation, depression, and less likely: post-traumatic stress disorder and obsessive compulsive disorder. Anxiety disorders and hyperventilation may produce chest pain and a positive family history of CAD may increase the individual's anxiety creating a vicious cycle. Anxiety can appear with a variety of symptoms including dyspnoea, palpitations, nausea and diaphoresis. As all of these symptoms are also characteristics of cardiac chest pain, efforts to obtain a definitive diagnosis may prove difficult.

Table 2.3 Signs and symptoms of depression.

- Anhedonia (lack of interest in activities, interests etc.)
- Low mood (including low self-esteem, crying, low self-worth)
- Altered sleeping patterns (especially early morning waking)
- Weight loss
- Increased use of alcohol, painkillers or sedatives
- Suicidal ideation

The role of depression is the development of CAD is now established and it is deemed an independent predictor of CAD. There is a paucity of studies examining prognosis in depressed patients with non-cardiac chest pain, however a recent study in those with heart failure demonstrated a poorer prognosis in those with worsening depression (Sherwood et al. 2011). The increased risk of adverse clinical outcomes suggest the need for depression screening in all patients who experience chest pain: whether it is eventually diagnosed as cardiac or non-cardiac, depressive symptoms need to be identified and treated appropriately.

Another clinical consideration in those with underlying psychological disorders is the presence of increased somatisation. Those with non-cardiac chest pain often report higher levels of severity of symptoms and more longstanding social problems (Mayou et al. 1999). Thus a thorough history of psychological and social status is important and needs to be considered in conjunction with the physical assessment findings and diagnostic tests in those who appear multiple times with recurrent chest pain.

INVESTIGATING CHEST PAIN

Once it has been established that the origin of the chest pain is not cardiac, patients should be reassured of this in oral and written forms and most importantly a cardiology consultation. Given the lack of a diagnosis in this cohort, further investigations to establish a diagnosis should be pursued. The worry of recurrent episodes of chest pain and the consequent lack of guidance can lead to depression and anxiety in individuals and a reduction in quality of life and ability to perform daily activities.

Unfortunately there is no quick method of distinguishing cardiac chest pain from non-cardiac chest pain and there is the risk that a small number of patients with undiagnosed ACS will be discharged. Approximately 50 to 60% of chest pain presentations in the ED will have no major ECG abnormalities and no history of CAD. There is evidence that those not admitted to hospital with non-cardiac chest pain have a 2 to 3 fold subsequent CAD morbidity and mortality risk compared to those admitted to hospital (Pope et al. 2000). This risk must be weighed up with the findings of other studies which have revealed high percentages of musculoskeletal conditions (25%), gastro-intestinal disorders (between 9 to 25%) (Karlson et al. 1991, Rouan et al. 1987, Katz et al. 1987).

A logistical problem that often occurs in clinical practice is the inability for patients to undertake an exercise treadmill test (also known as exercise ECG or treadmill stress testing). If a 12-lead ECG is inconclusive, the physician may decide an exercise test is appropriate. However the exercise test is contra-indicated in a vast range of pathologies including chronic obstructive pulmonary disease, obesity, peripheral arterial disease and degenerative conditions such as arthritis. More recently, there has been a move to investigate the presence of myocardial ischemia using stress echocardiography, myocardial perfusion scintigraphy and coronary CT angiography. An important factor about these sophisticated tests is the associated costs and this was alluded to in a recent editorial (Gerber et al. 2010).

Patients need to be assessed for the suitability to undertake an exercise ECG and it is a simple cost-effective method to determine the presence of CAD. Guidelines are available on undertaking exercise testing in ED chest pain centres and in those with unstable angina (Braunwald et al. 2002). Recent evidence demonstrates the benefit of coronary artery calcium score (CACS) to identify patients who can be discharged home without the need for further testing (Nabi et al. 2010). Two patients with a zero CACS (of 2,000 patients) were reported to have cardiac events during hospitalization or after discharge (0.3%) supporting the view that CACS has a role in identifying low risk patients and allows safe discharge with no further testing required.

A study of 300 patients who were deemed low to intermediate risk of ACS underwent thallium exercise ECGs prior to discharge from ED (Rahman et al. 2010). Of the 300 patients, two patients had positive Troponin results at 6 hr after chest pain onset and three patients had abnormal stress testing results (1%) and were subsequently identified as high-risk of ACS. The study highlighted that if accurate cardiovascular risk assessment is performed, those with positive results would have been identified using the guidelines and secondly that stress testing is not necessary prior to discharge in those deemed low risk. A one-yr follow-up of these patients revealed that 35 had further chest pain, three had MIs, four were diagnosed with angina and five patients had atrial fibrillation (see Fig. 2.1). A total of 75 specialist referrals were made: 30 patients had cardiology referrals and 20 patients had gastroenterology referrals. This one-yr follow-up highlights the ongoing healthcare utilization in this cohort deemed low to intermediate ACS risk and the fact that a small number of patients proceeded to have cardiac events (Lee et al. 2010).

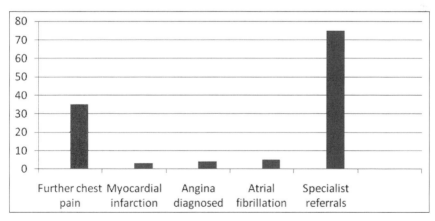

Figure 2.1 One year follow-up status of patients with low to intermediate of Acute Coronary Syndrome.

If chest pain is deemed non-cardiac, further investigations may be required and although many studies have reported high numbers of individuals with GI complaints, there is evidence that relatively few GI consultations and investigations are organized (Leise et al. 2010). GI diagnoses include reflux, dyspepsia, peptic ulcer disease and oesophageal spasm and the authors also noted the presence of CV risk factors including hypertension and hypercholesterolemia supporting the hypothesis that often angina is present with other co-morbidities.

In summary, in those with suspected angina establishing the cause is the priority and where positive tests are reported including ECG abnormalities and raised Troponin, the cardiologists need to decide on the need for urgent angiography or fibrinolysis. In those deemed intermediate to low-risk of ACS, the decision needs to be taken about whether the patient can be discharged home or remain an in-patient so that myocardial perfusion studies can be performed or referrals to other specialties made. Another alternative for managing patients is the use of rapid access clinics.

Rapid Access Clinics

With the increased ED presentations for chest pain, hospitals have opened rapid access chest pain clinics, chest pain observational units and short stay units. The purpose of these specialty units is to assess those who have experienced chest pain, undertake diagnostics tests to determine the cause and for those with confirmed ACS, offer rapid treatment usually fibrinolysis and percutaneous coronary interventions. The rapid access clinic seems to be a cost-effective and efficient method of providing care for this group of patients with low rates of subsequent cardiac events (Khan et al. 2010).

With the proliferation of rapid access chest pain clinics and the increased prevalence in chest pain presentations (with many proving to be non-cardiac), it is important to investigate healthcare utilization in those who were diagnosed with non-cardiac chest pain. This should include mortality and morbidity, subsequent consultations, primary and secondary providers, and medication use. As part of the assessment of non-cardiac chest pain and where a psychogenic cause is suspected, it is probably worth undertaking a selection of psychological questionnaires on anxiety and depression such as Hospital Anxiety and Depression Scale (HADS) and the Beck Depression Inventory. It is probably worth consulting with the psychiatry department on which tests are the most appropriate for the patient cohort.

One of the issues with psychogenic presentations is increased use of healthcare and multiple referrals for further tests etc. In one study with those who had persistent disabling chest pain, cognitive behavioural therapy (CBT) was offered (Mayou et al. 1999). A total of 133 patients were recruited with 69 having normal coronary angiograms and 64 not offered angiography but reassured that their chest pain was not cardiac in nature. At six wk follow-up, a small number of patients (n=56) continued to have clinically significant symptoms and a small percentage of these (15%) required intensive collaborative management. This research demonstrated two things: firstly that those who were given reassurance and no angiography tended to have longer histories of chest pain, more often reported breathlessness on exertion and had more emergency department admissions. Secondly, that there is a need for stepped aftercare encompassing appointments with the cardiologist offering written and verbal reassurance to individually tailored psychological treatment. However although psychological interventions and CBT may be indicated, not all patients may be willing to participate.

Differentiating between cardiac and gastrointestinal problems can be challenging and in those with both these conditions, it is necessary to educate the patient as much as possible so that they can appropriately manage these conditions when symptoms occur. One study identified that those with non-cardiac chest pain had higher rates of GI disorders in the year before diagnosis (Odds Ratio =2.0, 95% CI 1.5 to 2.7%) or dyspepsia (OR=1.7, 95% CI 1.4–2.2%) and were significantly associated with a diagnosis of non-cardiac chest pain (Ruigomez et al. 2009).

KEY FACTS ON GASTROINTESTINAL REFLUX DISORDER (GORD)

- Up to 60% of non-cardiac chest pain are diagnosed with GORD.
- GORD patients are four and half times more likely to be diagnosed in those with non-cardiac chest pain than controls.

- GORD is a significant co-morbidity in patients with non-cardiac chest pain.
- Significant levels of cardiovascular risk factors are seen in GORD patients.

Although chest pain may be determined non-cardiac, healthcare professionals may still wish to suggest lifestyle modification to individuals. This strategic long-term approach will benefit the individual and may prevent cardiovascular disease and indeed, cancer. Recommendations such as smoking cessation, increasing physical activity, weight management (especially in those who are overweight and obese) and optimal diabetes management should be undertaken regardless of specialty (see Table 2.4 below). Specifically in those with cardiovascular risk factors such as hypertension and raised cholesterol, lifestyle changes and pharmacological treatment should be discussed and importantly managed to ensure the optimal blood pressure and lipid profile is achieved. These lifestyle behaviours are associated with a myriad of chronic conditions and primary and secondary prevention will improve health in the short and long-term.

Table 2.4 Risk factor management.

Risk factor	Treatment recommendations
Smoking	-Aim to quit smoking
Raised cholesterol & triglycerides	- Prescribe statins unless contra-indicated - Increase omega 3 consumption - Increase consumption of oily fish or supplement diet with fish oil
Raised blood pressure	- Prescribe medication as required - Reduce salt intake (if overweight, encourage increased physical activity and dietary adjustments) AIM: BP <130/80mmHg
Poor diet	- Educate regarding diet and encourage a healthy diet (salt, saturated fats and alcohol reduction for example) - Refer to dietician etc as appropriate - Set realistic weight reduction goals
Sedentary lifestyle	- Encourage physical activity of 30 min per day and refer to cardiac rehabilitation, physiotherapy as required

MANAGING NON-CARDIAC CHEST PAIN

Some might argue that those with non-cardiac chest pain once reassured and discharged will not return but unfortunately this is not evident in practice. One study has shown that in non-cardiac chest pain, symptoms persist two yr after the initial presentation (Jain et al. 1997) and at higher rates than those with diagnosed CAD. One study reporting that up to 80% of patients who experienced chest pain in primary care did not have a diagnosis one yr after their initial consultation for non-cardiac chest pain (Ruigomez et al. 2009). These findings strongly support the need for adequate and appropriate follow-up and providing a diagnosis where possible.

RECOMMENDATIONS FOR MANAGING NON-CARDIAC PATIENTS

- Cardiovascular risk assessment should be performed on ALL patients regardless of the origin of their chest pain
- Out-patient cardiologist appointment to provide reassurance and written confirmation of the diagnosis (i.e., not cardiac disease) including a clear explanation of the symptoms and appropriate advice
- Offer follow-up as appropriate and referrals (including gastroenterology, respiratory etc.)
- Patients should be educated so they have a good understanding of their symptoms and
- All patients should be given a plan of care for dealing with ongoing symptoms.
- The GP should be informed of all diagnostic test results and be involved in the patient's care.

For those with anxiety and/or depressive symptoms:
- Consider stepped aftercare if symptoms persist
- Ensure collaborative multidisciplinary approach is implemented
- Intensive individual psychological therapy as required
- Consider specialist psychiatric treatment as necessary and/or cognitive behavioural therapy
- Organise regular follow-up at 2 wk, two, three and six mon as required

There is no doubt that appropriate healthcare is needed in the treatment of non-cardiac chest pain. There is a role for a specialist cardiac nurse to manage patients with non-cardiac chest pain with an emphasis on assessing the patient, especially for anxiety and depressive symptoms, psycho-social issues and providing education. The Australian National Heart Foundation has added an information sheet to their website about chest pain (both cardiac and non-cardiac) (National Heart Foundation 2010). An excellent reference is available from the European Society of Cardiology Committee for Practice Guidelines with a section on coronary heart disease and risk stratification is described in detail along with recommendations for treatment based on the findings (European Society of Cardiology 2010).

Co-ordination of patient care between the ED, cardiac services, rapid access clinics and a referral process for non-cardiac chest pain is important. A guideline that is user friendly should be implemented so that any healthcare professional can identify which tests and investigations completed and what, if any referrals or follow-up need to be arranged. Written information explaining symptoms about non-cardiac chest pain should be available to patients.

KEY POINTS

- Angina pectoris and chest pain presentations can be challenging to treat.
- The main priority is to relieve symptoms, identify those who have ACS and instigate treatment urgently.
- For those who have non-cardiac chest pain, appropriate follow-up and investigations should be undertaken as often there is an underlying pathology which is causing the symptoms.
- Where available, rapid access clinics should be used.
- Patients with non-cardiac chest pain require reassurance and written information with a plan of care.
- A significant number of patients have cardiovascular risk factors and these require optimal treatment and ongoing management.

In conclusion, there is now evidence to demonstrate that there is increasing prevalence of angina and chest pain and with increased healthcare costs, unnecessary investigations should be avoided and where-ever possible a diagnosis and clear explanation given to the patient. Cardiovascular risk factors are often present in these patients and although the chest pain may not be cardiac in origin, optimal management of these risk factors should be initiated. Chest pain remains a major healthcare issue and with collaborative processes between hospitals and the local GP and community facilities, the prevalence of chest pain presentations may be treated with better outcomes.

SUMMARY

- Angina pectoris is an early manifestation of coronary artery disease and requires urgent investigation to determine the cause.
- When results for angina are negative, a diagnosis of 'non-cardiac chest pain' is given. Some individuals have further diagnostic tests and are given gastro-intestinal, respiratory or musculoskeletal diagnosis.
- However, for some individuals, despite undergoing several tests, no diagnosis is made and this lack of a definitive diagnosis can lead to significant healthcare utilization.
- Many hospitals now offer rapid access clinics using a multi-disciplinary approach to manage chest pain.
- Recommendations in the management of non-cardiac chest pain that can be applied in clinical practice include undertaking anxiety and depression screening and performing a cardiovascular risk assessment.
- Reduction of saturated fats and alcohol can be very beneficial for those with gastro-intestinal complaints as well as reducing overall cardiovascular disease risk.
- Angina pectoris remains a prevalent condition and in those who are experience it for the first time, thorough investigations and where possible a diagnosis should be given. If the cause is determined to be non-cardiac, adequate follow-up and advice is required, appropriate specialist referrals made and where cardiovascular risk factors are identified, optimal management and monitoring is essential to prevent coronary artery disease.

DEFINITIONS

ACS: Acute coronary syndrome is a collective term for angina and cardiac presentations including myocardial infarctions.

CAD: Coronary Artery Disease is the result of atherosclerosis and increases the risk of heart attacks and angina.

GORD: Gastro-oesophageal reflux disorder is the term given to gastric symptoms that include indigestion and abdominal pain. Symptoms are often mistaken for angina.

MI: Myocardial Infarction is the result of a blockage in a coronary artery leading to death of the heart muscle (myocardium). It requires urgent medical attention.

LIST OF ABBREVIATIONS

ACS	:	Acute Coronary Syndrome
CAD	:	Coronary Artery Disease
GORD	:	Gastro-oesophageal reflux disorder
MI	:	Myocardial Infarction

REFERENCES CITED

Braunwald, E., E.M. Antman, J.W. Beasley, D. Melvin, M.D. Cheitlin, S. Judith, J.S. Hochman, R.H. Jones, D. Kereiakes, J. Kupersmith, T.N. Levin, C.J. Pepine, J.W. Schaeffer, E.E. Smith, D.E. Steward, P. Theroux, R.J. Gibbons, E.M. Antman, J.S. Alpert, D.P. Faxon, V. Fuster, G. Gregoratos, L.F. Hiratzka, A.K. Jacobs and S.C. Smith. 2002. ACC/AHA 2002 guideline update for the management of patients with unstable angina and non-ST-segment elevation myocardial infarction--summary article: a report of the American College of Cardiology/ American Heart Association task force on practice guidelines (Committee on the Management of Patients With Unstable Angina). J Am Coll Cardiol. Oct 2 2002; 40(7): 1366–74.

European Society of Cardiology Committee for Practice Guidelines. 2010. European Society of Cardiology Guidelines Desk Reference. 2010. Springer Healthcare. London, UK.

Gerber, T.C., M.C. Kontos and B. Kantor. 2010. Emergency department assessment of acute-onset chest pain: contemporary approaches and their consequences. Mayo Clin Proc 85(4): 309–313.

Ghezeljeh, T.N., M. Momtahen, M.K. Tessma, M.Y. Nikravesh, I. Ekman and A. Emami. 2010. Gender specific variations in the description, intensity and location of Angina Pectoris: A cross-sectional study. Int J Nurs Stud 47: 965–974.

Jain, D., D. Fluck, J.W. Sayers, S. Ray, E.A. Paul and A.D. Timmis. 1997. One step chest pain clinic can identify high cardiac risk. J R Coll Physician London 31: 401–404.

Karlson, B.W., J. Herlitz, P. Pettersson, H.E. Ekvall and A. Hjalmarson. 1991. Patients admitted to the emergency room with symptoms indicative of acute myocardial infarction. J Intern Med 230: 251–258.

Katz, P.O., C.B. Dalton, J.E. Richter, W.C. Wu and D.O. Castell. 1987. Esophageal testing of patients with noncardiac chest pain or dysphagia. Ann Intern Med 106: 593–597.

Khan, M.M., S. Mahida, V. Watson, C. Preston and D.S. Bulugahapitiya. 2010. Cardiac event rates in patients discharged from nurse-led rapid access chest pain clinic as having non-cardiac chest pain following initial triage. Eur J Cardiovasc Nurs 9S: S1.

Kotecha, D., M. Flather, M. McGrady, J. Pepper, G. New, H. Krum and D. Eccleston. 2010. Contemporary predictors of coronary artery disease in patients referred for angiography. Eur J Cardiovasc Prev Rehab 17: 280–288.

Lee, G., S. Dix, D. Mitra, J. Coleridge and P. Cameron. 2010. The efficacy and safety of a chest pain protocol for short stay unit patients—A 1-year follow-up. Austral Emerg Nurs J 13(4): 143.

Leise, M.D., R. Locke, R.A. Dierkhising, A.R. Zinsmeister, G.S. Reeder and N.J. Talley. 2010. Patients dismissed from the hospital with a diagnosis of noncardiac chest pain: cardiac outcomes and health care utilisation. Mayo Clin Pro 85(4): 323–330.

Mayou, R., B. Bryant, C. Forfar and D. Clark. 1994. Non-cardiac chest pain and benign palpitations in the cardiac clinic. Br Heart 72: 548–453.

Mayou, R., C.m. Bass and B.M. Byrant. 1999. Management of non-cardiac chest pain: from research to clinical practice. Heart 81: 387–392.

Nabi, F., S.M. Chang, C.M. Pratt, J. Paranilam, L.E. Peterson, M.E. Frias and J.J. Mahmarian. 2010. Coronary artery calcium scoring in the emergency department: Identifying which patients with chest pain can be safely discharged home. Ann Emerg Med 2010; 56: 220–229.

National Heart Foundation. 2010. Managing chest pain. http://www.heartfoundation.org.au/Heart_Information/Heart_Conditions/Chest_Pain/Managing_Chest_Pain/Pages/default.aspx.

National Institute for Health and Clinical Excellence. 2010. Chest pain of recent onset Assessment and diagnosis of recent onset chest pain or discomfort of suspected cardiac origin http://www.nice.org.uk/nicemedia/live/12947/47918/47918.pdf. ISBN 978-1-84936-187-3 March 2010.

Phan, A., C. Shufelt and C.M.B. Merz. 2009. Persistent chest pain and no obstructive coronary artery disease. JAMA 301(14): 1468–1474.

Pope, J.H., T.P. Aufderheide, R. Ruthazer, R.H. Woolard, J.A. Feldman, J.R. Beshansky, J.L. Griffith and H.P. Selker. 2000. Missed diagnoses of acute cardiac ischemia in the emergency department. New Engl J Med 2000; 342: 1163–70.

Rahman, F., B. Mitra, P.A. Cameron and J. Coleridge. 2010. Stress testing before discharge is not required for patients with low and intermediate risk of acute coronary syndrome after emergency department short stay assessment. Emerg Med Aust 5: 449–456.

Robertson, N., N. Javed, N.J. Samani and K. Khunti. 2008. Psychological morbidity and illness appraisals of patients with cardiac and non-cardiac chest pain attending a rapid access chest pain clinic: a longitudinal cohort study. Heart 2008. Doi: 10.1136/hrt.2006.100537.

Rosengren, A. 2008. Psychology of chest pain. Heart 266–267.

Rouan, G.W., J.R. Hedges, R. Toltzis, B. Goldstein-Wayne, D. Brand and L. Goldman. AffiliationsDepartment of Internal Medicine, University of Cincinnati Medical Center, Cincinnati, Ohio, USA. Department of Emergency Medicine, University of Cincinnati Medical Center, Cincinnati, Ohio, USA. 1987. A chest pain clinic to improve the follow-up of patients released from an urban university teaching hospital emergency department. Ann Emerg Med 16: 1145–1150.

Ruigomez, A., E.L. Masso-Gonzalez, S. Johansson, M. Wallander and L.A. Garcia-Rodriguez. 2009. Chest pain without established ischaemic heart disease in primary care patients: associated comorbidities and mortality. Br J Gen Prac e78–e86.

Russell, M., M. Williams, E. May and S. Stewart. 2010. The conundrum of detecting stable angina pectoris in the community setting. Nat Rev Cardiol 7: 106–113.

Sherwood, A., J.A. Blumenthal, A.L. Hinderliter, G.G. Koch, K.F. Adams, C.S. Dupree, D.R. Benimhon, K.S. Johnson, R. Trivedi, M. Bowers, R.H. Christenson and C.M. O'Connor. 2011. Worsening depressive symptoms are associated with adverse clinical outcomes in patients with heart failure. J Am Coll Cardiol 57: 418–423.

Coronary Plaque Rupture in Patients with Acute Coronary Syndrome

Toshio Imanishi

ABSTRACT

Coronary artery thrombosis superimposed on a disrupted atherosclerotic plaque initiates abrupt arterial occlusion and is the proximate event responsible for 60–70% causes of acute coronary syndrome. Recent studies have shown that plaque composition rather than plaque size or stenosis severity, has been recognized as a critical role in plaque rupture and thrombosis. Plaque rupture is defined as a necrotic core with a thin fibrous cap that is disrupted or ruptured, allowing the flowing blood to come in contact with the thrombogenic necrotic core. The exposure of tissue factor and other factors within the lipid-rich necrotic core is probably involved in the induction of thrombosis. The necrotic lipid core and plaque inflammation appear to be key factors. Extracellular matrix loss in the fibrous cap, a preclude to rupture, is attributed to matrix degrading enzymes as well as death of matrix synthesizing smooth muscle cells; inflammation appears to play a critical role in both these processes. Inflammatory cell derived tissue factor is also a key contributor to plaque thrombogenicity. Further detailed understanding of mechanisms causing plaque rupture should provide novel insights into prevention of athero-thrombotic cardiovascular events. This chapter provides a concise update on the evolving concepts in the pathophysiology of plaque rupture and thrombosis.

Imanishi, Department of Cardiovascular Medicine, Wakayama, Medical University, 811-1 Kimiidera, Wakayama 641-8510, Japan; Email: t-imani@wakayama-med.ac.jp

List of abbreviations after the text.

INTRODUCTION

Despite considerable therapeutic advances over the past 50 yr, cardiovascular disease is the leading cause of death worldwide. This is mainly a result of the increasing prevalence of atherosclerosis, owing to the ageing population, the improved survival of patients with atherosclerotic cardiovascular disease and, above all, the widespread under-recognition and under-treatment of individuals with risk factors for atherosclerosis. Atherosclerosis is characterized by the thickening of the arterial wall to form an atherosclerotic plaque, a process in which cholesterol deposition, inflammation, extracellular-matrix formation and thrombosis have important roles. The process of inflammation regulates atherosclerosis. In particular, macrophage has emerged as the key cellular mediator of inflammation in atheroma, and participates in all phases of atherogenesis, including lesion initiation, progression, and complication. Initially recruited to the artery wall as monocytes, these cells mature into macrophages (Fig. 3.1). Once resident, macrophages ingest oxidized lipoproteins via scavenger receptors, becoming foam cells, and contribute to atheroma expansion. A subset of macrophages can die within the atheromavia apoptosis (programmed cell death), leading to the development of a necrotic lipid core. Autopsy studies have demonstrated prominent macrophage accumulation in ruptured atherosclerotic lesions. These findings underscore a key role for macrophages in plaque complications. Plaque rupture is the most common type of plaque complication, according to about 70% of fatal acute myocardial infarction (AMI) and/or sudden coronary death. Several retrospective autopsy series and a few cross-sectional clinical studies have suggested that thrombotic coronary death and acute coronary syndrome (ACS) are caused by the plaque features and associated factors presented in Table 3.1.

The purpose of this chapter is to summarize the current knowledge regarding the coronary plaque rupture in patients with ACS.

PATHOPHYSIOLOGY OF CULPRIT LESION IN ACS

In vitro pathological studiesdemonstrated that most cases of ACS are thought to result from sudden luminal narrowing caused by

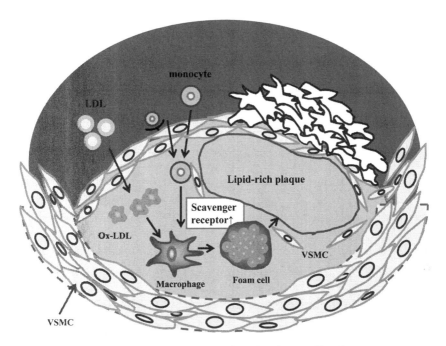

Figure 3.1 The evolution of the atherosclerotic plaque. Blood monocytes enter the arterial wall in response to chemoattractant cytokines such as monocyte chemoattracnt protein 1. Scavenger receptors mediate the uptake of modified lipoprotein particles and promote the development of foam cells. Atherosclerosis progresses through lipid core expansion and macrophage accumulation at the edges of the plaque, leading to fibrous cap rupture.

Color image of this figure appears in the color plate section at the end of the book.

Table 3.1 Criteria for defining vulnerable plaques [1].

Major criteria
1. Active inflammation (monocyte/macrophage and sometimes T-cell infiltration)
2. Thin cap with large lipid core
3. Endothelial denudation with superficial platelet aggregation
4. Fissured plaque
5. Stenosis > 90%

Minor criteria
1. Superficial calcified nodule
2. Glistening ywllow
3. Intraplaque hemorrhage
4. Endothelial dysfunction
5. Outward (positive) remodeling

thrombosis based on plaque rupture, erosion, and superficial calcified nodule (Naghavi et al. 2003, Virmani et al. 2005). Within these three different pathologies, plaque rupture has been reported to be the most frequent, plaque erosion to be the second and superficial calcified nodule to be the least (Naghavi et al. 2003, Virmani et al. 2005).Plaque rupture is defined as a lesion consisting of a large necrotic core with ruptured thin fibrous cap with overlying thrombus, and thin-cap fibroatheroma (TCFAs) without rupture are thought to be precursor lesions of ACS in pathology(Virmani et al. 2005). Plaque erosion demonstrates intra-luminal thrombus without discontinuity of the thick fibrous cap even in cases with necrotic core, although it is mostly devoid of a necrotic core (Naghavi et al. 2003, Virmani et al. 2005). Superficial calcified nodule, the least frequent cause of intracoronary thrombus, reveals the discontinuity of fibrous cap without endothelial cells by bony calcified nodule with overlying thrombus. Based on these three different features of plaque morphology in the culprit lesion in ACS, vulnerable plaques, which are defined as plaques prone to disruption and to be culprit plaques with thrombus, are proposed by five major and five minor criteria as listed in Table 3.1 (Naghavi et al. 2003). Major criteria for detection of a vulnerable plaque is as follows. (1) *Active inflammation*: Plaque with active inflammation may be identified by extensive macrophage accumulation. (2) *Thin cap with large lipid core*: These plaques have a cap thickness <100µm and a lipid core accounting for >40% of plaque's total volume. (3) *Endothelial denudation with superficial aggregation*: These plaques are characterized by superficial erosion and platelet aggregation or fibrin deposition. (4) *Fissured/injured plaque*: Plaques with a fissured cap (most of them involving a recent rupture) that did not result in occlusive thrombi may be prone to subsequent thrombosis, entailing occlusive thrombi or thromboemboli. (5) *Severe stenosis*: On the surface of plaques with severe stenosis, shear stress imposes a significant risk of thrombosis and sudden occlusion.

CHARACTERISTICS OF VULNERABLE PLAQUE

The plaque prone to rupture, that is vulnerable plaque, is made of a large lipid core composed of foam cells, apoptotic and necrotic cells, and debris, and is separated from the lumen by a fibrous cap (mainly comprising collagen, proteoglycans, and smooth muscle cells). A necrotic core characterizes plaque rupture with an overlying thin-ruptured cap infiltrated by macrophages. Smooth muscle cells (SMCs) within the cap are absent or few.

Plaque rupture occurs in the plaque fissuring at one point, which ultimately brings the platelets into contact with the content of the lipid core, and the blood coagulation factors together with tissue factor. Thus, the thinner the fibrous cap, the higher the risk of rupture. It has been postulated that TCFA is the precursor lesion of plaque rupture. TCFA was defined as a lesion with a fibrous cap <65μm thick and infiltrated macrophages (>25 cells per 0.3 mm diameter field) (Burke et al. 1997). A thickness of 65μm was chosen as a criterion of instability because, in rupture, the thickness of the fibrous cap near the rupture site measures 23 ± 19 μm, with 95% of the caps measuring <65μm (Burke et al. 1999).

MECHANISMS OF PLAQUE RUPTURE

The spectrum of clinical syndromes caused by coronary atherosclerosis ranges from asymptomatic disease and stable angina pectoris (SAP) to acute coronary syndrome (ACS), including unstable angina pectoris (UAP). ACS develops as a series of nonlinear events in an otherwise slowly progressive process. In fact, in 60-80% of patients with ACS, a coronary angiogram performed weeks or months before the acute event had shown the culprit site to have under 70% (often under 50%) diameter narrowing (Fig. 3.2) (Shah 2003). The nonlinearity has been attributed to a combination of factors, of which plaque rupture and superimposed thrombosis are considered the most important. Although the pathways leading to plaque rupture remain incompletely understood, recent research has shown that the participation of immune cells and inflammation is a key factor.

The concept of plaque rupture is based on the weakening of the fibrous cap which protects the blood from the thrombogenic lipid core. The extracellular matrix is actively destroyed by matrix metalloproteinases (MMPs), which are locally over-expressed by the macrophages (Shah et al. 1995, Shah 1998). In the weakened plaque, the number of macrophages producing MMP is increased and the number of smooth muscle cells (SMCs) repairing the extracellular matrix is decreased. This imbalance between extra-cellular matrix synthesis and degradation is a solid basis for plaque rupture.

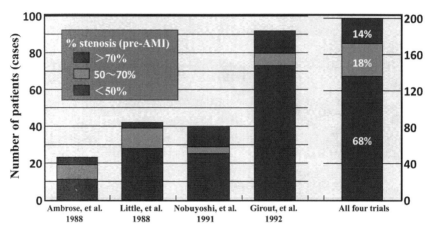

Figure 3.2 Relationship of plaque rupture to plaque size and stenosis severity. Retrospective analysis of serial angiograms, as well as prospective serial angiographic observations, have suggested that a coronary angiograms performed weeks or months before the acute event had shown the culprit site to have under 70% (often under 50%) diameter narrowing. Thus, plaques producing non-flow limiting and less than severe stenosis account for more cases and less than severe stenoses account for more cases of plaque rupture and thrombosis than plaques producing a more severe luminal diameter stenosis.(Adapted from Falk E, et al. Circulation 1995; 92: 657–71.)

In addition, activated T lymphocytes produce interferon γ, which decreases the synthesis of collagen I and III by SMCs and also influence the degradation of the extracellular matrix by stimulating macrophage production of MMP involving CD40 ligation.

VULNERABLE PLAQUES ASSESSED BY INVASIVE AND NON-INVASIVE MODALITIES

Although angiography still serves as a key test in the management of symptomatic coronary artery disease (CAD), several clinical observations have emphasized the need for a more detailed analysis of the structure and biology of atherosclerotic plaques. First, epidemiologic observations have shown that most patients who suffer a sudden cardiac event (acute ischemic syndromes or sudden cardiac death) have no prior symptoms. Second, it has been found that ACS often results from plaque rupture at sites with no or only modest luminal narrowing at an angiography. Therefore, there is considerable demand for diagnostic procedures that go beyond assessment of the vessel lumen to identify rupture-prone vulnerable plaques. The goal of these techniques is to identify plaques with regard to specific criteria of vulnerability, such as a thin fibrous cap and a large lipid core.

Intravascular Ultrasound (IVUS)

Cross sectional IVUS images of normal coronary arteries demonstrate a sharp endothelial border, thin, clear echo-lucent media, and dense adventitial layer (Nishimura et al. 1990). When intimal thickening occurs during development of atherosclerotic plaques, signal free zones represent lipid accumulations, soft echoes correspond to SMCs, fibrosis or collagen deposition, and bright intensity indicates calcification (Wickline 2004, Honda et al. 2004). Further rupture or ulceration of lesions can often be detected in culprit vessels (Maehara et al. 2002). An IVUS interrogation of all three epicardial vessels in more than 200 patients closely concurred with the pathologic literature that plaque rupture of the infarct-related vessel occurred in two-thirds of patients representing AMI; of these, one-fifth of the patients had evidence of multiple ruptures (Hong et al. 2004). The plaques underlying culprit lesions were hypoechoic, eccentric, and positively remodeled. In fact, lesions that develop into an acute coronary event were characteristically large and contained relatively shallow but prominent echo-lucent zones (Yamaguchi et al. 2000). These features support the concept that vulnerable plaques have a large necrotic core and thin fibrous caps, and demonstrate outward (positive) remodeling at the site of culprit lesions.

Optical Coherence Tomography (OCT)

Optical coherence tomography (OCT) is an optical analogue of IVUS, which provides high-resolution images of the coronary arteries up to 10μm. OCT has the capacity to reveal the coronary lesion morphologies in detail including types of plaque, and plaque rupture (Fig. 3.3). OCT also allows us to demonstrate the microstructures of an atherosclerotic plaque, including TCFA. OCT-derived TCFA is defined as a plaque with lipid content in> 2 quadrants and the

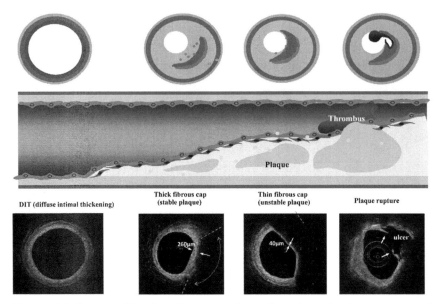

Figure 3.3 A generalized and progressive atherothrombosis process and corresponding optical coherence tomographic images. (Upper) This 'time-line' depiction of the development of atherothrombosis in a coronary vessel emphasizes the long-term nature of this process. (lower) Optical coherence tomography (OCT) provides high-resolution images of the coronary artery up to 10 μm.

Color image of this figure appears in the color plate section at the end of the book.

thinnest part of the fibrous cap measuring <70μm (Fig. 3.4). Frequency of TCFA in the initial and recent reports was 65 and 83%, respectively, and the thickness of fibroucs cap in the former and latter studies was 47 and 49±21μm, respectively (Jang et al. 2005, Kubo et al. 2007). There were no significant differences in these characteristics between men and women (Chia et al. 2007). More interestingly, fibrous cap thickness and the site of rupture have been reported to be different in the type of ACS onset in a recent study(Tanaka et al. 2008a), and it is thinner (50μm vs. 90μm, respectively) in rest-onset ACS compared with exertion-triggered ACS, and the rupture at plaque shoulder is more frequent in the latter (57 vs. 93%). Furthermore, not only TCFA but also thick cap (up to 150μm) fibroatheroma (ThCFA) might have a chance to rupture, and high-sensitive C-reactive protein level negatively correlates well in proportion to the thickness of the ruptured fibrous cap. These findings suggest that various types of disruption should occur in similar frequency even *in vivo* as reported

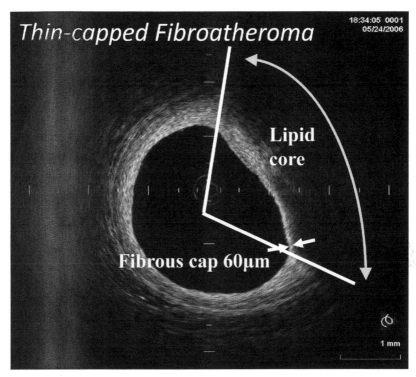

Figure 3.4 OCT-derived TCFA. A fibrous cap was identified as a signal-rich homogenous region overlying a lipid core, which was characterized by a signal-poor region on the OCT image. The lesion with a fibrous cap of <65μm was diagnozed as OCT-derived TCFA.

Color image of this figure appears in the color plate section at the end of the book.

in vitro histology studies, and fibrous cap rupture could be expected at the site of plaque shoulder even in ThCFA up to 150μm during exertion. Exercise-induced high shear stress at the site of the plaque shoulder might be speculated to be a cause of fibrous cap disruption, although the demonstration of inflammation by commercially available OCT systems has not been established yet. However, the possibility of macrophage accumulation by OCT has been already described in the culprit lesion in ACS (Tearney et al. 2003, MacNeill et al. 2004).

It has been reported recently that the frequency of plaque rupture (43% vs. 13% vs. 71%, $p<0.001$) and plaque ulceration (32% vs. 7% vs. 8%, $p=0.003$) are significantly different among the types of UAP in Braunwald class I, II, and III, and the fibrous cap thickness (140μm vs. 150μm vs. 60μm, $p<0.001$), minimal lumen area (0.70 mm^2, vs. 1.80mm^2 vs. 2.31mm^2, $p<0.001$), and the frequency of thrombus (72% vs. 30% vs. 73%, $p<0.001$) are also significantly different among the types of UAP, although lipid-rich plaque is observed frequently in each group (85% vs. 80% vs. 87%, $p=0.73$) (Mizukoshi et al. 2010). Although there has been no report in the difference of rupture portions in the culprit plaques by OCT as reported in a previous longitudinal IVUS analysis between AMI and UAP (Tanaka et al. 2008b), class III UAP might be more vulnerable than the other classes because plaque rupture in TCFA overlying thrombus is more frequent even in enough of the lumen area. In class I, organic luminal narrowing and plaque ulceration with small amount thrombus might be a mechanism of ACS. Further comparative study among ACSs in a large number would be required to demonstrate the differences in the culprit lesion morphology in detail in ACS.

Multidetector Computed Tomography (MDCT)

Recently, developed multidetector computed tomography (MDCT), particularly 64-slice MDCT, is a robust and reliable non-invasive imaging modality that permits visualization of the coronary artery wall and has been shown to have a high diagnostic accuracy for the detection and quantification of vulnerable plaque properties (Fig. 3.5) (Motoyama et al. 2009, Kashiwagi et al. 2009). Several studies have demonstrated CT characteristics plaques associated with acute coronary syndrome (ACS), including positive remodeling (PR), low CT attenuationplaques (LAPs), and spotty calcification. Motoyama et al. reported that localization of plaques showing both PR and LAPs portended a higher likelihood of ACS development over the next two yr (Motoyama et al. 2009). Kashiwagi et al. found that plaques with PR and LAPs are MDCT-identified morphologic features, similar to OCT-identified TCFA (Kashiwagi et al. 2009).

PLAQUE STABILIZATION BY STATINS

One of the most effective medical therapies to prevent a coronary event at the moment is lipid-lowering therapy using 3-hydroxy-3-methylglutaryl coenzyme A reductase inhibitors (statins). Several

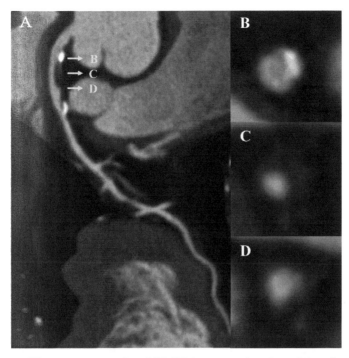

Figure 3.5 The representative MDCT images of vulnerable plaques. A representative image of vulnerable coronary plaques assessed by multidetector computed tomography (MDCT) is shown. Panel A shows curved mutiplanar reformation images of right coronary artery (RCA). Positive remodeling, low-attenuation plaque, and spotty calcification were detected in RCA #1 on MDCT. Panels B, C, and D show the cross sectional images of vulnerable plaque sites.

clinical trials have suggested statins could stabilize vulnerable plaques of coronary artery and prevent ischemic events. Although precise mechanisms for efficacy of statins treatment remain unclear, the treatment may result in stabilization of vulnerable through increase in the thickness of fibrous-cap and reduction of inflammation in patients with vulnerable plaque (Fig. 3.6). MIRACL study indicated that treatment with statins, initiated during the acute phase of ACS, reduces the risk of early, recurrent ischemic events. These results demonstrated the effects of aggressive low density lipoprotein (LDL)-lowering therapy to stabilize vulnerable plaques rapidly, and therefore statins have a good indication to treat coronary atherosclerosis even for secondary prevention of ACS. In the PROVE IT-TIMI 22 study of 3,745 patients with ACS, Ridker reported that those with LDL-C <70 mg/dl or CRP <2 mg/l after statin therapy

Scientific Basis of Healthcare: Angina

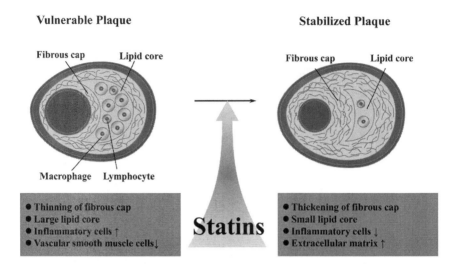

Figure 3.6 Effects of statins on plaque vulnerability. The pathological features of the most common form of vulnerable plaque include a large lipid pool within the plaque, a thin fibrous cap, and macrophage accumulation within the cap. Statins affect coronary plaque vulnerability, that is, the changes from vulnerable to stable plaque, which increase fibrous cap thickness and a decrease in total atheroma volume. (Adapted from Libby P, et al. Nat Med 2002; 8: 1257–62.)

Color image of this figure appears in the color plate section at the end of the book.

had a lower rate of recurrent coronary events compared with those with LDL-C or CRP that remained higher than these levels (Ridker et al. 2001). Furthermore, rosuvastatin significantly reduced the incidence of major cardiovascular event even in healthy subjects without hyperlipidemia but with elevated high sensitive CRP. Our group also showed that statins have the potential to regress plaque volume and increase the tension of the fibrous cap in vulnerable plaques by reducing both LDL and CRP, respectively (Takarada et al. 2010).

PRACTICE AND PROCEDURES (PRACTICAL METHODS AND GUIDELINES)

The Evolution of the Atherosclerotic Plaque

According to a simplified modification of the criteria set for the by the American Heart Association (AHA) Committee on Vascular lesions

(Stary et al. 1995), plaque progression can be subdivided into five pathologically/clinically relevant phases.

Phase 1 (early). Lesions are small, commonly seen in young people, and categorized into three types as follows: type I lesions consisting of macrophage-derived foam cells that contain lipid droplets; type II lesions, consisting of both macrophages and SMCs and mild extracellular lipid droplets; and type III lesions, consisting of SMCs surrounded by extracellular connective tissueand lipid deposits.

Phase 2 (advanced). Lesions, although not necessarily stenotic, may be prone to rupture because of their high lipid content, increased inflammation, and thin fibrous cap. These plaques are categorized morphologically as one of two variants: type IV lesions, consisting of confluent cellular lesions with a great deal of extracellular lipid intermixed with normal intima, which may predominate as an outer layer or cap; or type Va lesions, processing an extracellular lipid core covered by an acquired fibrous cap. Phase 2 plaques can evolve into the acute phases 3 and 4.

Phase 3. These lesions are characterized by acute complicated type IV lesions, originating from ruptured (type IV or Va) or eroded lesions, and leading to mural, non-obstructive thrombosis. This process is clinically silent, but occasionally may lead to the onset of angina.

Phase 4. The lesions are characterized by acute complicated type VI lesions, with fixed or repetitive occlusive thrombosis. This process becomes clinically apparent in the form of an ACS, although not infrequently it is silent. About two-thirds of ACS are caused by occlusive thrombosis on a non-stenotic plaque, although in about one-third, the thrombus occurs on the surface of a stenotic plaque. In phase 3 and 4, changes in the geometry of ruptured plaques, as well as organization of the occlusive or mural thrombus by connective tissue, can lead to the occlusive or significantly stenotic and fibrotic plaques.

Phase 5. These lesions are characterized b type Vb (calcific) or Vc (fibrotic) lesions that may cause angina; however, if preceded by stenosis or occlusion with associated ischemia, the myocardium may be protected by collateral circulation and such lesions may then be silent.

KEY FACTS

An Etiologic Approach for Unstable Angina

- Five pathophysiological processes may contribute to the development of ACS.
- Plaque rupture or erosion with superimposed nonocclusive thrombus.
- Dynamic obstruction (i.e., coronary spasm of an epicardial artery).
- Progressive mechanical obstruction.
- Inflammation.
- Secondary unstable angina, related to increased myocardial oxygen demand or decreased supply (e.g., anemia).

PATHOLOGICAL CHARACTERISTICS OF RUPTURED PLAQUE

The pathological features of the most common form of vulnerable plaques include a large lipid pool inside a plaque, a thin fibrous cap, and macrophage accumulation within the cap, resulting in the expression of proteolytic enzymes that weaken the fibrous cap and ultimately promote plaque rupture. Even though the plaque is large, the lumen is not significantly obstructed because of the expansive (or outward) remodeling of the vessel. Such plaques are often associated with adventitial proliferation of vasa vasorum and neovascularization of the lesion.

- Plaque morphology
    Large necrotic core: >25% of plaque area
    Vessel remodeling
    Large plaque size
    Neovascularization
    Intraplaque hemorrhage
- Thin fibrous cap (<65μm)
- Foam cells in fibrous cap
    Increased apoptosis (macrophage, vascular smooth muscle cells)
    Increase matrix metalloproteinases

SUMMARY POINTS

- Plaque rupture is the most common type of plaque complication, accounting for about 70% of fatal acute myocardial infarction and/or sudden cardiac deaths.
- ACS develops as a series of nonlinear events in an otherwise slowly progressive process.
- It is likely that both inflammation and immune cells play a key role for plaque rupture.
- The concept of plaque is based on the weakening of the fibrous cap which protects the blood from the thrombogenic lipid core.
- The imbalance between extra-cellular matrix synthesis and degradation is a solid basis for plaque rupture.
- The plaque prone to rupture, that is vulnerable plaque, is made of a large lipid core composed of foam cells, apoptotic and necrotic cells, and debris.
- OCT allows us to demonstrate clearly culprit lesion morphologies in detail including types of plaque, plaque disruption, thrombus, fibrous cap thickness, and the frequency of thin-capped fibroatheroma (TCFA).
- TCFA, which is a precursor lesion of plaque rupture, is defined a lesion with a fibrous cap <65μm thick and infiltrated macrophages (>25 cells per 0.3 mm diameter field).
- The thinner the fibrous cap, the higher the risk of rupture.
- It is important to identify patients in whom disruption of a vulnerable plaque is likely to result in a clinical event.
- Most of patients with ACS, a coronary angiogram performed weeks or months before the acute event had shown the culprit site to have under 70% (often under 50%) diameter narrowing.
- Available screening and diagnostic methods are insufficient to identify the victims before the event occurs.
- Ruptured plaques showed extensive macrophage infiltration of the fibrous cap, in particular at rupture sites contrary to stable lesions, which contained fewer inflammatory cells.
- Statins have the potential to regress plaque volume and increase the tension of the fibrous cap in vulnerable plaques.

DEFINITIONS

Atherosclerosis: Inflammatory disease that occurs preferentially in patients with risk factors and a genetic predisposition.

Non-atherosclerotic intimal thickening: The normal accumulation of smooth muscle cells (SMCs) in the intima in the absence of lipid or macrophage foam cells.

Fatty streak: Luminal accumulation of foam cells without a necrotic core or fibrous cap.

Atherosclerotic intimal thickening: SMCs in a proteoglycan-rich matrix with area of extracellular lipid accumulation without necrosis.

Fibrous cap atheroma: Well formed necrotic core with an overlying fibrous cap.

Culprit lesion: A lesion in a coronary artery considered to be responsible for the clinical event.

High-risk, vulnerable, and thrombosis-prone plaque: These terms can be used to synonyms to describe a plaque that is at increased risk of thrombosis and rapid stenosis progression.

Coronary plaque rupture: It is defined as a lesion consisting of a large necrotic core with ruptured thin fibrous cap with overlying thrombosis.

Coronary plaque erosion: It is defined as intra-luminal thrombus without discontinuity of the thick fibrous cap.

Optical coherence tomography (OCT): is a recently developed optical imaging technique that provides high-resolution, cross-sectional images of tissue *in situ*.

LIST OF ABBREVIATIONS

ACS	:	acute coronary syndrome
AMI	:	acute myocardial infarction
CAD	:	coronary artery disease
CRP	:	C-reactive protein
IVUS	:	intravascular ultrasound
LAPs	:	low CT attenuation plaques
LDL	:	low density lipoprotein
MDCT	:	multidetector computed tomography

MMPs	:	matrix metalloproteinases
OCT	:	optical coherence tomography
Ox-LDL	:	oxidized-low density lipoprotein
TCFA	:	thin cap fibroatheroma
ThCFA	:	thick cap fibroatheroma
PR	:	positive remodeling
SAP	:	stable angina pectoris
SMCs	:	smooth muscle cells
Statins	:	3-hydroxy-3-methylglutaryl coenzyme A reductase inhibitors
UAP	:	unstable angina pectoris

REFERENCES CITED

Burke, A.P., A. Farb, G.T. Malcom, Y.H. Liang, J. Smialek and R. Virmani. 1997. Coronary risk factors and plaque morphology in men with coronary disease who died suddenly. N Engl J Med vol. 336, pp. 1276–1282.

Burke, A.P., A. Farb, G.T. Malcom, Y. Liang, J.E. Smialek and R. Virmani. 1999. Plaque rupture and sudden death related to exertion in men with coronary artery disease. JAMA vol. 281, pp. 921–926.

Chia, S., O.C. Raffel, M. Takano, G.J. Tearney, B.E. Bouma and J.K. Jang. 2007. In-vivo comparison of coronary plaque characteristics using optical coherence tomography in women vs. men with acute coronary syndrome. Coronary Artery Dis vol. 18, pp. 423–7.

Honda, O., S. Sugiyama, K. Kugiyama, H. Fukushima, S. Nakamura, S. Koide, S. Kojima, N. Hirai, H. Kawano, H. Soejima, T. Sakamoto, M. Yoshimura and H. Ogawa. 2004. Echolucent carotid plaques predict future coronary events in patients with coronary artery disease. J Am Coll Cardiol vol. 43, pp. 1177–1184.

Hong, M.K., G.S. Mintz, C.W. Lee, Y.H. Kim, S.W. Lee, J.M. Song, K.H. Han, D.H. Kang, J.K. Song, J.J. Kim, S.W. Park and S.J. Park. 2004.Comparison of coronary plaque rupture between stable angina and acute myocardial infarction: a three vessel intravascular ultrasound study in 235 patients. Circulation vol. 110, pp. 928–933.

Jang, I.K., G.J. Tearney, B. MacNeill, M. Takano, F. Moselewski, N. Iftima, M. Shishkov, S. Houser, H.T. Aretz, E.F. Halpern and B.E. Bouma. 2005. In vivo characterization of coronary atherosclerotic plaque by use of optical coherence tomography. Circulation vol. 111, pp. 1551–1555.

Kashiwagi, M., A. Tanaka, H. Kitabata, H. Tsujioka, H. Kataiwa, K. Komukai, T. Tanimoto, K. Takemoto, S. Takarada, T. Kubo, K. Hirata, N. Nakamura, M. Mizukoshi, T. Imanishi and T. Akasaka. 2009. Non-invasive assessment of thin-cap fibroatheroma by multidtetector computed tomography. JACC. Cardiovasc. Imaging vol. 2, pp. 1412–19.

Kubo, T., T. Imanishi, S. Takarada, A. Kuroi, S. Ueno, T. Yamano, T. Tanimoto, Y. Matsuo, T. Masho, H. Kitabata, K. Tsuda, Y. Tomobuchi and T. Akasaka. 2007. Assessment of culprit lesion morphology in acutemyocardial infarction: ability of optical coherence tomography compared with intravascular ultrasound and coronary angioscopy. J Am Coll Cardiol vol. 50, pp. 933–9.

MacNeill, B.D., I.K. Jang, B.E. Bouma, N. Iftimia, M. Takano, H. Yabushita, M. Shishkov, C.R. Kauffman, S.L. Houser, H.T. Aretz, D. DeJoseph, E.F. Halpern and G.J. Tearney. 2004. Focal and multi-focal plaque macrophage distributiuons in patients with acute and stable presentations of coronary artery disease. J Am Coll Cardiol vol. 44, pp. 972–9.

Maehara, A., G.S. Mintz, A.B. Bui, O.R. Walter, M.T. Castagna, D. Canos, A.D. Pichard, L.F. Satler, R. Waksman, W.O. Suddath, J.R. Laird Jr, K.M. Kent and N.J. Weissman. 2002. Morphologic and angiographic features of coronary plaque rupture detected by intravascular ultrasound. J Am Coll Cardiol vol. 40, pp. 904–910.

Mizukoshi, M., T. Imanishi, A. Tanaka, T. Kubo, Y. Liu, S. Takarada, H. Kitabata, T. Tanimoto, K. Komukai, K. Ishibashi and T. Akasaka. 2010. Clinical classification and plaque morphology determined by optical coherence tomography in unstable angina pectoris. Am J Cardiol vol. 106, pp. 323–28.

Motoyama, S., M. Sarai, H. Harigaya, H. Anno, K. Inoue, T. Hara, H. Naruse, J. Ishii, H. Hishida, N.D. Wong, R. Virmani, T. Kondo, Y. Ozaki and J. Narula. 2009.Computed tomographic angiography characteristics of atherosclerotic plaques subsequently resulting in acute coronary syndrome. J Am Coll Cardiol vol. 54, pp. 49–57.

Naghavi, M., P. Libby, E. Falk, S.W. Casscells, S. Litovsky, J. Rumberger, J.J. Badimon, C. Stefanadis, P. Moreno, G. Pasterkamp, Z. Fayad, P.H. Stone, S. Waxman, P. Raggi, M. Madjid, A. Zarrabi, A. Burke, C. Yuan, P.J. Fitzgerald, D.S. Siscovick, C.L. de Korte, M. Aikawa, K.E. Juhani Airaksinen, G. Assmann, C.R. Becker, J.H. Chesebro, A. Farb, Z.S. Galis, C. Jackson, I.K. Jang, W. Koenig, R.A. Lodder, K. March, J. Demirovic, M. Navab, S.G. Priori, M.D. Rekhter, R. Bahr, S.M. Grundy, R. Mehran, A. Colombo, E. Boerwinkle, C. Ballantyne, W. Insull Jr, R.S. Schwartz, R. Vogel, P.W. Serruys, G.K. Hansson, D.P. Faxon, S. Kaul, H. Drexler, P. Greenland, J.E. Muller, R. Virmani, P.M. Ridker, D.P. Zipes, P.K. Shah and J.T. Willerson. 2003. From vulnerable plaque to vulnerable patient: a call for new definitions and risk assessment strategies. Part I: Circulation vol. 108, pp. 1664–1672.

Nishimura, R.A., W.D. Edwards and C.A. Warnes. 1990. Intravascular ultrasound imaging *in vitro* validation and pathologic correlation. J Am Coll Cardiol vol. 16, pp. 145–154.

Ridker, P.M., N. Rifai, M. Clearfield, J.R. Downs, S.E. Weis, J.S. Miles, A.M. Gotto Jr and Air Force/Texas Coronary Atherosclerosis Prevention Study Investigators. 2001. Measurement of C-reactive protein for the targeting of statin therapy in the primary prevention of acute coronary events. N Engl J Med vol. 344, pp. 1959–1965.

Shah, P.K. 1998. Role of inflammation and metalloproteinases in plaque disruption and thrombosis. Vasc Med vol. 3, pp. 199–206.

Shah, P.K. 2003. Mechanisms of plaque vulnerability and ruptures. J Am Coll Cardiol vol. 41, pp. 15S–22S.

Shah, P.K., E. Falk, J.J. Badimon, A. Fernandez-Ortiz, A. Mailhac, G. Villareal-Levy, J.T. Fallon, J. Regnstrom and V. Fuster. 1995. Human monocyte-derived macrophages induce collagen breakdown in fibrous caps of atherosclerotic plaques. Potential role of matrix-degrading metalloproteinases and implications for plaque rupture. Circulation vol. 92, pp. 1565–1569.

Stary, H.C., A.B. Chandler, R.E. Dinsmore, V. Fuster, S. Glagov, W. Insull Jr, M.E. Rosenfeld, C.J. Schwartz, W.D. Wagner and R.W. Wissler. 1995. A definition of advanced types of atherosclerotic lesions and a histologic classification of atherosclerosis. A report from the Committee on Vascular Lesions of the Council on Atherosclerosis. American Heart Association. Circulation vol. 92, pp. 1355–74.

Takarada, S., T. Imanishi, K. Ishibashi, T. Tanimoto, K. Komukai, Y. Ino, H. Kitabata, T. Kubo, A. Tanaka, K. Kimura, M. Mizukoshi and T. Akasaka. 2010. The effect of lipid and inflammatory profiles on the morphological changes of lipid-rich plaques in patients with non-ST elevated acute coronary syndrome: follow-up study optical coherence tomography and intra-vascular ultrasound. JACC. Cardiovasc. Interv vol. 3, pp. 766–772.

Tanaka, A., T. Imanishi, H. Kitabata, T. Kubo, S. Takarada, T. Tanimoto, A. Kuroi, H. Tsujioka, H. Ikejima, S. Ueno, H. Kataiwa, K. Okouchi, M. Kashiwaghi, H. Matsumoto, K. Takemoto, N. Nakamura, K. Hirata, M. Mizukoshi and T. Akasaka. 2008a. Morphology of exertion-triggered plaque rupture in patients with acute coronary syndrome: an optical coherence tomography. Circulation vol. 118, pp. 2368–73.

Tanaka, A., K. Shimada, M. Namba, T. Sakamoto, Y. Nakamura, Y. Nishida, J. Yoshikawa and T. Akasaka. 2008b. Relationship between longitudinal morphology of ruptured plaques and TIMI flow grade in acute coronary syndrome: a three-dimensional intravascular ultrasound imaging study. Eur Heart J vol. 29, pp. 38–44.

Tearney, G.J., H. Yabushita, S.L. Houser, H.T. Aretz, I.K. Jang, K.H. Schlendorf, C.R. Kauffman, M. Shishkov, E.F. Halpern and B.E. Bouma. 2003. Quantification of macrophage content in atherosclerotic plaques by optical coherence tomography. Circulation vol. 107, pp. 113–9.

Virmani, R., F.D. Kolodgie, A.P. Burke, A.V. Finn, H.K. Gold, T.N. Tulenko, S.P. Wrenn and J. Narula. 2005. Atherosclerotic plaque progression and vulnerability to rupture: angiogenesis as a sourse of intraplaque hemorrhage. Arterioscler. Thromb Vasc Biol vol. 25, pp. 2054–2061.

Waters, D.D., G.G. Schwartz, A.G. Olsson, A. Zeiher, M.F. Oliver, P. Ganz, M. Ezekowitz, B.R. Chaitman, S.J. Leslie, T. Stern and MIRACL Study Investigators. Schwartz. 2001. Myocardial ischemia reduction with aggressive cholesterol lowering (MIRACL) study investigators. Effects of atrovastatin onearly recruitment ischemic events in acute coronary syndromes: the MIRACL study: a randomized controlled trial. JAMA vol. 285, pp. 1811–1718.

Wickline, S.A. 2004. Plaque characterization surrogate markers or the real thing? J. Am. Coll. Cardiol. vol. 43, pp. 1185–87.

Yamagishi, M., M. Terashima, K. Awano, M. Kijima, S. Nakatani, S. Daikoku, K. Ito, Y. Yasumura and K. Miyatake. 2000. Morphology of vulnerable coronary plaque: insights from follow-up of patients examined by intravascular ultrasound before an acute coronary syndrome. J Am Coll Cardiol vol. 35, pp. 106–111.

4

The Use and Application of Angina Questionnaires

Norul Badriah Hassan[1],* and Muhammad Termizi Hassan[2]

ABSTRACT

Studies have reported that patients with angina are at higher risks of having poor quality of life and fatal cardiovascular events. Early detection and intervention are essential to prevent or delay progression to more serious manifestations of coronary artery disease and cardiovascular death. Several standardized questionnaires with different aims, psychometric properties, strengths and limitations are available for use in angina. For the detection of stable angina pectoris, the Rose Angina Questionnaire is commonly used. For the measurement of quality of life in angina, several generic and disease-specific questionnaires are available. The most widely used generic quality of life instruments in angina are the Medical Outcomes Study 36-Item Short Form Health Survey (SF-36) and Nottingham Health Profile Questionnaire. For the disease- specific questionnaire, the Seattle Angina Questionnaire is extensively used. Combination of several questionnaires has also been used in many angina studies. Choice for an

---

[1]Lecturer, Pharmacology Department, School of Medical Sciences, Universiti Sains Malaysia (USM), Kampus Kesihatan, Kubang Kerian, 16150, Kelantan, Malaysia; Email: norul@kb.usm.my
[2]Emergency Medicine Specialist Registrar, Emergency Department, The Adelaide and Meath Hospital, Incorporating the National Children's Hospital, Tallaght, Dublin 24, Ireland; Email: termizihassan@gmail.com
*Corresponding author

List of abbreviations after the text.

appropriate questionnaire should be based on a careful consideration of the instrument's psychometric properties, depth and breadth of relevant questions or domains, study aims and expected outcomes.

USE AND APPLICATION OF ANGINA QUESTIONNAIRES

Studies have reported that in many individuals who are at risk of major cardiovascular events, prevalence of undetected stable angina pectoris (SAP) is high (Kones 2010) or suboptimally managed (Russell et al. 2010). Identification of angina is very important since undetected individuals with the disease are at higher risk of having poor quality of life (QOL) and fatal cardiovascular events. Measuring QOL in angina is also essential since it provides a more valid description of actual treatment effects from the patients' perspective. Early detection and appropriate intervention can delay progression to more serious manifestations of coronary artery disease (CAD) and cardiovascular death.

Typical angina symptoms are usually characterized by retrospinal chest pain, usually described as heaviness or pressure, radiating to the ulnar aspect of the left arm, neck, jaw, or shoulders. Chest discomfort is caused by transient myocardial ischemia without necrosis. Episodes of exertional angina may last from 2–10 min and are relieved by rest within 1–5 min, or more rapidly with sublingual nitroglycerin.

Several methods are available for angina detection or diagnosis (Rose 1962), assessment of patient's belief (Furze et al. 2003), patient preference for treatment (Bowling et al. 2010), work limitations (Lerner et al. 1998) and measurement of health-related QOL outcomes in angina (Ware and Sherbourne 1992, Spertus et al. 1995). These methods differ relatively in terms of their aims, validity, reliability, strengths and limitations. Choice for an appropriate method for a study should be based on a careful consideration of the instrument's psychometric properties, depth and breadth of relevant questions or domains, specific purposes and expected outcomes.

Self-report using questionnaires are frequently preferred over other detection methods since they are simpler, practical, non-invasive, faster and cheaper. This approach is widely used, and especially so in epidemiological studies. A questionnaire is an indirect and convenient method to study prevalence, incidence, frequency, distributions and outcomes of angina. A clinically useful questionnaire should be acceptable to health care professionals, practical to administer and scientifically sound in terms of basic psychometric properties such as validity, reliability, and responsiveness.

For the detection and assessment of typical symptoms of angina in general population, Rose Angina Questionnaire (RAQ) (Rose 1962) is widely used. For the measurement of health outcomes or QOL in angina, several validated questionnaires are available. These include generic questionnaires (Table 4.1) that provide multiple aspects of overall health (Ware and Sherbourne 1992) and disease-specific questionnaires solely developed for use in angina (Marquis et al. 1995, Spertus et al. 1995). Generic instruments address a wide range of domains and may be used for various diseases to determine how treatments affect daily activities, well-being and social functioning in the broader impact. On the other hand, disease-specific instruments are assumed to have the advantages of being more sensitive to detect changes in QOL. Different types of questionnaires have their relative strengths and weaknesses which justify researchers to combine several instruments in their studies (Kim et al. 2005). This chapter will focus on four of the most widely used questionnaires in angina.

Table 4.1 Disease-Specific and Generic Questionnaires for the Measurement of Health-Related Quality of Life in Stable Angina Pectoris.

Disease-Specific Questionnaire	Generic Questionnaire
Seattle Angina Questionnaire	Medical Outcomes Study 36-Item Short Form Health Survey (SF-36)
Quality of Life after Myocardial Infarction (QLMI/ MacNew Heart Disease) Questionnaire	Nottingham Health Profile Questionnaire (NHP) Part I
Angina Pectoris Quality of Life Questionnaire	Short form-12 (SF-12)
	Quality Well Being index (QWB)
	Sickness Impact Profile (SIP)
	Disease Impact Profile
	Quality of Life Index-Cardiac Version III

Unpublished

PRACTICE AND PROCEDURES: THE USE AND APPLICATION OF ANGINA QUESTIONNAIRES

The Rose Angina Questionnaire (RAQ)

The RAQ (Table 4.2) was developed in the 1960s and is also referred to as the London School of Hygiene Cardiovascular Questionnaire (Rose 1962). It is a standardized instrument for the screening of SAP and has been extensively used in cardiovascular epidemiology. Studies have used the RAQ in its original version (Cook et al. 1989), modified version (Table 4.3) (Lawlor et al. 2003), or combined with other measures (Najafi-Ghezeljeh et al. 2008). These versions can be used as self-administered questionnaires or for interview (Ford et al. 2000, Lawlor et al. 2003). The seven-item RAQ has been translated into several languages such as Spanish (Cosin et al. 1999), Thai (Udol and Mahanonda 2000), Bahasa Melayu (Hassan et al. 2007), Arabic (Moussa et al. 1994), Hindi, Bengali, Urdu and Punjabi (Fischbacher et al. 2001).

Definition and Classification of Rose Angina

Rose angina was confirmed (Rose 1962, Oei et al. 2004) with the presence of chest pain or discomfort plus the following four characteristics:

1) site must include *either* the sternum (any level) *or* the left anterior chest and the left arm.
2) it must be provoked by either hurrying or walking uphill or walking on the level.
3) when it occurs on walking, it must make the subject either stop or slacken pace, unless nitrates are taken.
4) it must disappear within 10 min from the time the subject stands still.

If one of the criteria was not met, angina pectoris was classified as absent. However, many authors (Cook et al. 1989, Lampe et al. 2001, Ugurlu et al. 2008) argued that the use of the standard WHO (Rose) definition of angina may underestimate symptom prevalence and further classified angina as definite or possible:

Table 4.2 WHO Rose Angina Questionnaire (RAQ).

---

1. Have you ever had any pain or discomfort or any pressure or heaviness in your chest?
   Yes   No
   If Yes, proceed to the next question
2. Do you get the pain in the centre of the chest or left chest or left arm?
   Yes   No
   If Yes, proceed to the next question
3. Do you get it when you walk at an ordinary pace on level or when you walk uphill or hurry?
   Yes   No
4. Do you slow down if you get the pain while working?
   Yes   No
5. Does the pain go way if you stand still or if you take a tablet under the tongue?
   Yes   No
6. Does the pain go away in less than 10 minutes?
   Yes   No
7. Have you ever had severe chest pain across the front of your chest lasting for half an hour or more?
   Yes   No

If the answer to question 3 or 4 or 5 or 6 or 7 is yes, patients may have angina and need referral. This questionnaire can be used as a screening tool for angina in epidemiological studies.
Adapted with permission from Hassan et al. 2007 (adapted from the WHO CVD risk management package for low and medium resource settings, 2002).
WHO: World Health Organization

Table 4.3 Shortened Version of the Rose Angina Questionnaire (RAQ).

---

1) Do you ever have any pain or discomfort in your chest?
   Yes   No
2) When you walk at an ordinary pace on the level does this produce the pain?
   Yes   No   Unable
3) When you walk uphill or hurry does this produce the pain?
   Yes   No   Unable

The shortened version of the Rose Angina Questionnaire focused on exertional chest pain. Adapted with permission from Lawlor et al. 2003.

a) *Definite angina* is defined as having pain or discomfort in the chest when walking uphill or hurrying and fulfilling all of the following four criteria:

  1) distribution of pain on specific location on the chest (e.g., sternum or left anterior chest with or without left arm and may vary, especially in women).

2) pain caused the participant to stop or slow down.
3) pain went away when the participant stopped or slowed down.
4) pain relieved within 10 min.

b) *Possible angina:* Chest pain was present on exertion but did not fulfil all four of the additional criteria.

Definite or possible angina was considered to be Rose angina (Cook et al. 1989). Angina can also be graded (Lawlor et al. 2003) as:

a) *Grade I angina (not severe):* if pain was brought on only by hurrying or walking uphill (No to Question 3)

b) *Grade II angina (severe):* if pain was brought on by walking at an ordinary pace on level ground (Yes to Question 3).

Validity of the RAQ has been extensively studied and debated. Positive Rose angina has been documented to predict ischemic heart disease in men (Lawlor et al. 2003) and mortality associated with CAD in men and women (LaCroix et al. 1990). Persistence of Rose angina symptoms was associated with severe disease and poor prognosis (Lampe et al. 2001), thicker carotid artery walls, greater amounts of cigarette smoking, greater prevalence of reported heart attack and greater use of chest pain medications (Sorlie et al. 1996). In women, Rose angina is associated with coronary risk factors (Nicholson et al. 1999), coronary calcification (Oei et al. 2004), resting ECG abnormalities (Nicholson et al. 1999) and carotid intima-media thickness (Sorlie et al. 1996).

The Seattle Angina Questionnaire (SAQ)

The Seattle Angina Questionnaire (SAQ) (Spertus et al. 1994; Spertus et al. 1995) is one of the most widely used specific measures to assess QOL outcomes in patients with angina (McGillion et al. 2008, Lu et al. 2009). The English version (SAQ- UK) (Garratt et al. 2001) has been specifically been modified and evaluated in British patients from general practice. A validated web-based version (Bliven et al. 2001) is also available to facilitate data administration and reporting. The SAQ has being translated in many languages such as German (Hofer et al. 2003), Russian (Syrkin et al. 2001), Norwegian (Pettersen et al. 2005) and Japanese (Seki et al. 2010). The 19-item SAQ (Spertus et al. 1995) measures five dimensions of CAD which includes:

1. *Anginal Stability (1-item):* measures whether patient's symptoms are changing over time.

2. *Anginal Frequency (2- item):* measures frequency of symptoms and use of medication.

3. *Physical Limitation (9-item):* measures how patient's daily activities are limited by angina symptoms.

4. *Treatment Satisfaction (4-item):* measures overall satisfaction and satisfaction with treatment and doctor explanations.

5. *Quality of Life (3-item):* measures overall impact of angina on patient's quality of life.

All items are measured on ordinal descriptive five or six point-Likert scales. Scores are calculated by summing the items within a dimension and converted to a 0–100 scale with one indicating the lowest or poorest response and 100 representing the best response. Higher scores indicate better levels of functioning (minimal physical limitation, less frequent angina, or better quality of life) except for the angina stability scale.

The angina stability scale requires a larger change to be clinically significant. A previous study has demonstrated that changes between 5 and 8 points are required to be clinically significant (Spertus et al. 1994). Generally, scores below 50 indicate a worsening of symptoms, scores of 50 indicate no change in angina symptoms, and scores above 50 indicate improvement. The SAQ requires only 5 min to be completed by patients.

The SAQ has well-established psychometric properties and has been used in many studies (Schroter and Lamping 2006, Lu et al. 2009). It is reported to be sensitive to clinical changes (Spertus et al. 1995), as responsive as the Coronary Revascularisation Outcome Questionnaire (Schroter and Lamping 2006), and able to predict mortality (Spertus et al. 2002). For the specific SAQ domains, physical limitation of the patients was affected by age, gender, severity of CAD (Lu et al. 2009) and presence of diabetes (Stone et al. 2008). The angina stability was affected by the history of myocardial infarction while the angina frequency was affected by the history of myocardial infarction and gender. For treatment satisfaction, it was affected by the severity of CAD while the disease perception was affected by the history of diabetes mellitus (Lu et al. 2009).

The Short Form 36 (SF-36)

The Medical Outcomes Study 36-Item Short Form Health Survey (SF-36) (Ware and Sherbourne 1992) is one the most widely used generic measure in QOL studies (Dougherty et al. 1998). The second improved version of SF-36 is available as software for easier data management and interpretation (Ware 2000). This questionnaire has been extensively validated and widely translated and adapted in more than 40 countries.

The SF-36 consists of 36 items with eight health domains. The domains include physical functioning, social functioning, role limitations due to physical health, role limitations due to mental health, general health, mental health, vitality, and bodily pain. The scales are scored and transformed to a 0–100, with higher score indicating higher levels of functioning or better health. It is suitable for self-administration, computerized administration or interview by trained personnel.

The SF-36 is extensively tested and commonly used in CAD studies (Schroter and Lamping 2006). Individuals with angina had higher increased risk of impaired physical functioning compared to those without angina and this effect was the same in men and women (Hemingway et al. 2003). All the SF-36 domains were significantly reduced with the presence of diabetes (Stone et al. 2008).

The Nottingham Health Profile Questionnaire (NHP)

The Nottingham Health Profile questionnaire (NHP) is a commonly used generic measure for QOL assessment in angina (Visser et al. 1994; Gandjour and Lauterbach 1999). It is widely used in the United Kingdom and several other countries (Peric et al. 2006).

The validated NHP consists of two parts. Part I comprised of 38 items and is categorized into six areas: energy, pain, emotional reactions, sleep, social isolation, and physical mobility. For Part II, it covers seven areas of daily life perceived as most affected by health: work, looking after the home, social life, home life, sex life, hobbies and interests, and vacations. A higher score indicates a higher level of dysfunction and worse QOL. Scores range from 0 to 100 and the higher the scores the greater the perceived dysfunction.

In a study using NHP Part I, worse results were observed in terms of physical mobility, social isolation and emotional reactions, lack of energy necessary for daily activities, and sleeping in patients with higher severity of angina (Peric et al. 2006).

Limitation of angina questionnaires

Generally, questioning patients may result in misrepresentation, bias, and overestimation. Questionnaires are solely based on patient recall, which may subject them to recall bias. Perception of previous symptoms may change considerably over time. Furthermore, a questionnaire developed in a different time, disease, country, or cultural context may not be a valid and reliable measure for another study population. Cross-cultural issues should be considered when a questionnaire is translated into different languages. However, many studies do not report the cross-cultural adaptation or translation process (Haywood et al. 1998, Baigi et al. 2001).

Determination of angina mainly based on chest pain alone is too broad to define SAP. Patients with CAD were diagnosed without cardiac chest pain in half of emergency admissions (Kones 2010). Pain may also arise from sources other than the heart (Colgan et al. 1988) and perceptions of pain may differ (Schofield et al. 1988). Chest pain may signify psychological morbidity rather than coronary disease (Colgan et al. 1988). Remission of symptoms may occur and this can affect pain assessment (Rosengren et al. 1986). Furthermore, tissue may no longer produce ischemic pain in the occurrence of myocardial infarction, changes in levels of activity, and effective anti-angina therapy.

Several authors reported variable sensitivity and specificity with the RAQ, especially in younger women (Garber et al. 1992, Sorlie et al. 1996). Association of Rose angina with cardiovascular risk markers is weaker in women than in men (Oei et al. 2004). In the absence of accepted gold standard in angina (Russell et al. 2010), measurement of validity can be affected. Gender differences in the presentation of chest angina pain, especially in younger females, are now recognized and acknowledged (Kones 2010).

Various RAQ versions with different definitions have been used and these may affect the results. Furthermore, the RAQ may be unreliable in different patient populations (Fischbacher et al. 2001).

Despite all these limitations, the RAQ is still an important method for the purpose of intervention and secondary prevention (Russell et al. 2010). However, diagnosis of SAP should be performed clinically and further supported by laboratory data.

The SAQ was not designed to discriminate patients with or without SAP (Russell et al. 2010). Since the SAQ is disease-specific, it is irrelevant in other disease states. For the generic SF-36 and NHP questionnaires, they may not examine therapy or related adverse effects in details. They may lack precision in measuring outcomes that are specifically relevant to SAP. These questionnaires require more time for completion and may be a burden to respondents. A shorter one page version of SF-12 is available and has been used in angina studies (Consoli et al. 2001).

Key Facts: Use and Application of Angina Questionnaires

1. Individuals with angina are at higher risk of having poor quality of life and fatal cardiovascular events.
2. Early detection and appropriate intervention can delay progression to more serious manifestations of coronary artery disease and cardiovascular death.
3. Several methods are available for the detection of angina, diagnosis, measurement of patient's belief, preference, and QOL.
4. The questionnaire approach is frequently preferred over other methods since it is simpler, practical, non-invasive, faster and cheaper.
5. The Rose Angina Questionnaire is commonly used for the prevalence detection and assessment of typical symptoms of SAP in general population.
6. For the assessment of QOL, disease- specific Seattle Angina Questionnaire is widely used.
7. As for generic measure of QOL, the Medical Outcomes Study 36-Item Short Form Health Survey (SF-36) and the Nottingham Health Profile questionnaire (NHP) are most commonly used.
8. Each questionnaire has its own strengths and limitations.
9. Choice for an appropriate questionnaire should be based on a careful consideration of the instrument's psychometric properties, depth and breadth of relevant questions or domains, aims and expected outcomes.

Key Facts: Advantages of Using Angina Questionnaires

1. Simpler
2. Practical
3. Non-invasive
4. Faster
5. Cheaper
6. Permit anonymity

KEY FACTS: PSYCHOMETRIC

- Branch of psychology
- Deals with design, administration, and interpretation of quantitative tests for the measurement of mental processes or functions.
- Applied widely in educational and psychological assessment.
- Concerned with the construction and validation of measurement instruments, such as questionnaires.
- Three basic psychometric properties are validity, reliability, and responsiveness.

SUMMARY

- Early detection and intervention are very important in angina to prevent or delay progression of serious manifestations of coronary artery disease (CAD) and cardiovascular death.
- Several standardized questionnaires with different aims, psychometric properties, strengths and limitations are available for use.
- The Rose Angina Questionnaire is widely used to measure the presence of angina in population studies.
- The Seattle Angina Questionnaire is a disease-specific questionnaire solely developed for use in angina.
- The Medical Outcomes Study 36-Item Short Form Health Survey (SF-36) and are the most widely used generic quality of life measures in angina.
- Combination of several questionnaires has also been used in many studies.

DEFINITIONS

Internal consistency: estimates reliability by grouping questions in a questionnaire that measure the same concept.

Likert scales: subjective scoring system that allows a person being surveyed to quantify likes and preferences, usually have 5-potential choices (strongly agree, agree, neutral, disagree, strongly disagree) or more.

Prevalence: proportion of individuals in a population having a disease. It is a statistical concept referring to the number of cases of a disease that are present in a particular population at a specific time.

Psychometry: branch of psychology that focuses on the design, administration, and interpretation of quantitative tests for the measurement of psychological variables such as intelligence, aptitude, and personality traits. It is primarily concerned with the construction, validity and reliability of the measurement instruments.

Quality of life: a term to measure sense of well-being and ability to carry out and enjoy various normal life activities. Health related quality of life encompasses domains of life directly affected by changes in health. Non-health-related quality of life includes features both the natural and created environment and personal resources. These factors affect health-related QOL, but are less likely to improve with appropriate medical care.

Questionnaire: a written or printed form which consists of a set of questions to be submitted to one or more persons used for gathering information on certain topics or subjects and making data comparable and easy for analysis.

Reliability: consistency or repeatability of a measurement, or the degree to which an instrument measures the same way each time it is used under the same conditions with the same subjects. A measure is considered reliable if a person's score on the same test given twice is similar.

Sensitivity: proportion of people with a disease who are correctly diagnosed (probability that the test is positive based on diagnostic criteria). The higher the sensitivity of a test or diagnostic criteria, the lower the rate of 'false negatives,' or people who have a disease but are not identified through the test.

Specificity: probability of a person who does not have a disease being correctly identified by a clinical test. It is expressed as the proportion of true negative results to the total of true negative and false positive results.

Symptom: sign of what the patient experiences about the illness, disease or injury. It can be a physical condition which shows that one has a particular illness or disorder or sensation or change in health function.

Validity: considers whether the questionnaires measure the concepts which they are supposed to measure.

LIST OF ABBREVIATIONS

CAD	:	coronary artery disease
NHP	:	Nottingham Health Profile questionnaire
QOL	:	quality of life
RAQ	:	Rose Angina Questionnaire
SAP	:	stable angina pectoris
SAQ	:	Seattle Angina Questionnaire
SF-36	:	Short Form 36 Questionnaire
WHO	:	World Health Organization

REFERENCES CITED

Baigi, A., B. Marklund and B. Fridlund. 2001. The association between socio-economic status and chest pain, focusing on self-rated health in a primary health care area of Sweden. Eur J Public Health 11: 420–424.

Bliven, B.D., S.E. Kaufman and J.A. Spertus. 2001. Electronic collection of health-related quality of life data: validity, time benefits, and patient preference. Qual Life Res 10: 15–22.

Bowling, A., B. Reeves and G. Rowe. 2010. The Patients Preferences Questionnaire for Angina treatment: results and psychometrics from 383 patients in primary care in England. Qual Saf Health Care 19: e9.

Colgan, S.M., P.M. Schofield, P.J. Whorwell, D.H. Bennett, N.H. Brooks and P.E. Jones. 1988. Angina-like chest pain: a joint medical and psychiatric investigation. Postgrad Med J 64: 743–746.

Consoli, S., L. Guize, P. Ducimetiere, I. Duprat-Lomon and I. Girod. 2001. Characteristics and predictive value of quality of life in a French cohort of angina patients. Arch Mal Coeur Vaiss 94: 1357–1366.

Cook, D., A. Shaper and P. MacFarlane. 1989. Using the WHO (Rose) angina questionnaire in cardiovascular epidemiology. Int J Epidemiol 18: 607–613.

Cosin, J., E. Asin, J. Marrugat, R. Elosua, F. Aros, M. de los Reyes, A. Castro-Beiras, A. Cabades, J.L. Diago, L. Lopez-Bescos, et al. 1999. Prevalence of angina pectoris in Spain. PANES Study group. Eur J Epidemiol 15: 323–330.

Dougherty, C.M., T. Dewhurst, W.P. Nichol and J. Spertus. 1998. Comparison of three quality of life instruments in stable angina pectoris: Seattle Angina Questionnaire, Short Form Health Survey (SF-36), and Quality of Life Index-Cardiac Version III. J Clin Epidemiol 51: 569–575.

Fischbacher, C.M., R. Bhopal, N. Unwin, M. White and K.G. Alberti. 2001. The performance of the Rose angina questionnaire in South Asian and European origin populations: a comparative study in Newcastle, UK. Int J Epidemiol 30: 1009–1016.

Ford, E.S., W.H. Giles and J.B. Croft. 2000. Prevalence of nonfatal coronary heart disease among American adults. Am Heart J 139: 371–377.

Furze, G., P. Bull, R.J. Lewin and D.R. Thompson. 2003. Development of the York Angina Beliefs Questionnaire. J Health Psychol 8: 307–315.

Gandjour, A., K.W. Lauterbach. 1999. Review of quality-of-life evaluations in patients with angina pectoris. Pharmacoeconomics 16: 141–152.

Garber, C.E., R.A. Carleton and G.V. Heller. 1992. Comparison of Rose Questionnaire Angina to exercise thallium scintigraphy: different findings in males and females. J Clin Epidemiol 45: 715–720.

Garratt, A.M., A. Hutchinson and I. Russell. 2001. The UK version of the Seattle Angina Questionnaire (SAQ-UK): reliability, validity and responsiveness. J Clin Epidemiol 54: 907–915.

Hassan, N.B., S.R. Choudhury, L. Naing, R.M. Conroy and A.R. Rahman. 2007. Inter-rater and intra-rater reliability of the Bahasa Melayu version of Rose Angina Questionnaire. Asia Pac J Public Health 19: 45–51.

Haywood, L.J., C. Faucett, M. deGuzman, K. Ell, S. Norris and E. Butts. 1998. Predictive value of prior Rose angina for myocardial infarction confirmation after emergency admissions. J Natl Med Assoc 90: 241–252.

Hemingway, H., M. Shipley, A. Britton, M. Page, P. Macfarlane and M. Marmot. 2003. Prognosis of angina with and without a diagnosis: 11 year follow up in the Whitehall II prospective cohort study. BMJ 327: 895.

Hofer, S., W. Benzer, G. Schussler, N. von Steinbuchel and N.B. Oldridge. 2003. Health-related quality of life in patients with coronary artery disease treated for angina: validity and reliability of German translations of two specific questionnaires. Qual Life Res 12: 199–212.

Kim, J.R., A. Henderson, S.J. Pocock, T. Clayton, M.J. Sculpher and K.A. Fox. 2005. Health-related quality of life after interventional or conservative strategy in patients with unstable angina or non-ST-segment elevation myocardial infarction: one-year results of the third Randomized Intervention Trial of unstable Angina (RITA-3). J Am Coll Cardiol 45: 221–228.

Kones, R. 2010. Recent advances in the management of chronic stable angina I: approach to the patient, diagnosis, pathophysiology, risk stratification, and gender disparities. Vasc Health Risk Manag 6: 635–656.

LaCroix, A.Z., J.M. Guralnik, J.D. Curb, R.B. Wallace, A.M. Ostfeld and C.H. Hennekens. 1990. Chest pain and coronary heart disease mortality among older men and women in three communities. Circulation 81: 437–446.

Lampe, F.C., P.H. Whincup, A.G. Shaper, S.G. Wannamethee, M. Walker and S. Ebrahim. 2001. Variability of angina symptoms and the risk of major ischemic heart disease events. Am J Epidemiol 153: 1173–1182.

Lawlor, D.A., J. Adamson and S. Ebrahim. 2003. Performance of the WHO Rose angina questionnaire in post-menopausal women: are all of the questions necessary? J Epidemiol Community Health 57: 538–541.

Lerner, D.J., B.C. Amick, 3rd, S. Malspeis, W.H. Rogers, D.R. Gomes and D.N. Salem. 1998. The Angina-related Limitations at Work Questionnaire. Qual Life Res 7: 23–32.

Lu, Y.H., L.X. Sun, J.H. Yan, Q. Li, Y.J. Wang, X. Zhuang, Z.F. Zhang and Z.J. Fan. 2009. Assessment of quality of life in patients with coronary artery disease with Seattle angina questionnaire. Zhonghua Yi Xue Za Zhi 89: 2827–2830.

Marquis, P., C. Fayol and J.E. Joire. 1995. Clinical validation of a quality of life questionnaire in angina pectoris patients. Eur Heart J 16: 1554–1560.

McGillion, M., H. Arthur, J.C. Victor, J. Watt-Watson and T. Cosman. 2008. Effectiveness of Psychoeducational Interventions for Improving Symptoms, Health-Related Quality of Life, and Psychological well Being in Patients with Stable Angina. Curr Cardiol Rev 4: 1–11.

Moussa, M.E., M. Gadallah and A.K. Mortagy. 1994. The evaluation of an Arabic version of Rose Questionnaire. J Egypt Public Health Assoc 69: 31–36.

Najafi-Ghezeljeh, T., I. Ekman, M.Y. Nikravesh and A. Emami. 2008. Adaptation and validation of the Iranian version of Angina Pectoris characteristics questionnaire. Int J Nurs Pract 14: 470–476.

Nicholson, A., I.R. White, P. Macfarlane, E. Brunner and M. Marmot. 1999. Rose questionnaire angina in younger men and women: gender differences in the relationship to cardiovascular risk factors and other reported symptoms. J Clin Epidemiol 52: 337–346.

Oei, H.H., R. Vliegenthart, J.W. Deckers, A. Hofman, M. Oudkerk and J.C. Witteman. 2004. The association of Rose questionnaire angina pectoris and coronary calcification in a general population: the Rotterdam Coronary Calcification Study. Ann Epidemiol 14: 431–436.

Peric, V.M., M.D. Borzanovic, R.V. Stolic, A.N. Jovanovic and S.R. Sovtic. 2006. Severity of angina as a predictor of quality of life changes six months after coronary artery bypass surgery. Ann Thorac Surg 81: 2115–2120.

Pettersen, K. I., A. Reikvam and K. Stavem. 2005. Reliability and validity of the Norwegian translation of the Seattle Angina Questionnaire following myocardial infarction. Qual Life Res 14: 883–889.

Rose, G. 1962. The diagnosis of ischaemic heart pain and intermittent claudication in field surveys. Bull World Health Organ 27: 645–658.

Rosengren, A., M. Hagman, K. Pennert and L. Wilhelmsen. 1986. Clinical course and symptomatology of angina pectoris in a population study. Acta Med Scand 220: 117–126.

Russell, M., M. Williams, E. May and S. Stewart. 2010. The conundrum of detecting stable angina pectoris in the community setting. Nat Rev Cardiol 7: 106–113.

Schofield, P.M., P.J. Whorwell, P.E. Jones, N.H. Brooks and D.H. Bennett. 1988. Differentiation of "esophageal" and "cardiac" chest pain. Am J Cardiol 62: 315–316.

Schroter, S., D.L. Lamping. 2006. Responsiveness of the coronary revascularisation outcome questionnaire compared with the SF-36 and Seattle Angina Questionnaire. Qual Life Res 15: 1069–1078.

Seki, S., N. Kato, N. Ito, K. Kinugawa, M. Ono, M. Motomura, A Yao, M Watanabe, Y Imai, N Takeda, M Inoue, M Hatano and K Kazuma. 2010. Validity and Reliability of Seattle Angina Questionnaire Japanese Version in Patients with Coronary Artery Disease. Asian Nursing Research 4: 57–63.

Sorlie, P.D., L. Cooper, P.J. Schreiner, W. Rosamond and M. Szklo. 1996. Repeatability and validity of the Rose questionnaire for angina pectoris in the Atherosclerosis Risk in Communities Study. J Clin Epidemiol 49: 719–725.

Spertus, J.A., J.A. Winder, T.A. Dewhurst, R.A. Deyo and S.D. Fihn. 1994. Monitoring the quality of life in patients with coronary artery disease. Am J Cardiol 74: 1240–1244.

Spertus, J.A., J.A. Winder, T.A. Dewhurst, R.A. Deyo, J. Prodzinski, M. McDonell and S.D. Fihn. 1995. Development and evaluation of the Seattle Angina Questionnaire: a new functional status measure for coronary artery disease. J Am Coll Cardiol 25: 333–341

Spertus, J.A., P. Jones, M. McDonell, V. Fan and S.D. Fihn. 2002. Health status predicts long-term outcome in outpatients with coronary disease. Circulation 106: 43–49.

Stone, M.A., K. Khunti, I. Squire and S. Paul. 2008. Impact of comorbid diabetes on quality of life and perception of angina pain in people with angina registered with general practitioners in the UK. Qual Life Res 17: 887–894.

Syrkin, A.L., E.A. Pechorin and S.V. Drinitsina. 2001. Validation of methods assessing the quality of life in patients with stable angina pectoris. Klin Med (Mosk) 79: 22–25.

Udol, K., N. Mahanonda. 2000. Comparison of the Thai version of the Rose questionnaire for angina pectoris with the exercise treadmill test. J Med Assoc Thai 83: 514–522.

Ugurlu, S., E. Seyahi and H. Yazici. 2008. Prevalence of angina, myocardial infarction and intermittent claudication assessed by Rose Questionnaire among patients with Behcet's syndrome. Rheumatology (Oxford) 47: 472–475.

Visser, M.C., A.E. Fletcher, G. Parr, A. Simpson and C.J. Bulpitt. 1994. A comparison of three quality of life instruments in subjects with angina pectoris: the Sickness Impact Profile, the Nottingham Health Profile, and the Quality of Well Being Scale. J Clin Epidemiol 47: 157–163.

Ware, J.E. Jr. 2000. SF-36 health survey update. Spine (Phila Pa 1976) 25: 3130–3139.

Ware, J.E. and C.D. Sherbourne. 1992. The MOS 36-item Short-Form Health Survey (SF-36). Medical Care 30: 473–483.

American College of Cardiology/American Heart Association Guidelines for ST-Elevation Myocardial Infarction—Overview of Guidelines for Care Before, During, and After STEMI

Rohit S Loomba[1,*] and *Rohit R Arora*[2]

ABSTRACT

ST-elevation myocardial infarction (STEMI) is one of the leading causes of mortality in hospitals today and, thus, it is vital that evidence based guidelines for STEMI management are continually revised and that healthcare workers are made aware of new changes. STEMI management begins with prevention and patient education and extends through the event itself and secondary prevention. Each step in the response to STEMI is equally important and contributes to reduction

---

[1]Chicago Medical School, Department of Cardiology, 3333 Green Bay Road, North Chicago, IL; Email: Loomba.rohit@gmail.com
[2]Captain James A Lovell Federal Health Center/Chicago Medical School, Department of Cardiology, 3001 Green Bay Road, North Chicago, IL; Email: Rohit.arora@va.gov
*Corresponding author

List of abbreviations after the text.

of resulting morbidity and mortality. Those involved in this response include the patient, the patient's family, emergency medical service, and physicians, among others. The multifaceted nature of the response to STEMI requires that there is proper response at each step in the process to ensure the most appropriate STEMI management. Additionally, initial steps involving patient and community response require proper counseling to see to it that initial response is appropriate. In this chapter we outline the current STEMI guidelines for primary prevention, patient recognition, community response, emergency medical system response, emergency department response, revascularization, and in-hospital treatment. We address responsibilities of bystanders to physicians in STEMI response, highlighting the importance of proper communication and understanding between various stages of response to minimize response times and inefficiency.

INTRODUCTION

ST-elevation myocardial infarction (STEMI) is one of the most common, potentially-fatal causes of hospitalization in the United States. It is important that physicians properly identify patients at risk for STEMI and educate them on recognition of symptoms and how to properly respond to these symptoms so that they can receive appropriate medical care. Many studies focus on the management of STEMI and it is important for physicians to be aware of the STEMI guidelines that have been developed and their updates so that there can be used to deliver the most appropriate care. This chapter focuses on the American College of Cardiology (ACC)/American Heart Association (AHA) 2004 guidelines for the management of STEMI as well as the subsequent 2007 and 2009 focused updates. Levels of evidence will be provided in parenthesis throughout (Table 5.1).

PRACTICE AND PROCEDURES: STEMI GUIDELINESMANAGEMENT BEFORE STEMI

Initial management of STEMI begins with the primary care physician who must properly identify the risk factors for coronary heart disease (CHD) present in the patient. Reassessment of risk factors should be done every 3 to 5 yr (I,C), keeping in mind that most of the risk is accounted for by gender, age, dyslipidemia, hypertension, diabetes, obesity, diet, and activity level (Yusuf et al. 2004) (Antman et al.

Table 5.1 Recommendation classes and levels of evidence.

Class I	Class IIa	Class IIb	Class III
Benefits outweigh risks and intervention should be used.	Benefits outweigh risks but there is still need for additional evidence. Intervention should be strongly considered.	Benefits at least equal to risks. Additional evidence still needed.	Risks outweigh benefits. Do not use intervention.
Level A	**Level B**	**Level C**	

Level A	Level B	Level C
Data from several randomized controlled trials or meta-analyses.	Evidence from limited number of randomized trials or nonrandomized evaluations.	Evidence from case studies but no randomized or nonrandomized studies.

Original table by Loomba et al. Description of the various recommendation classes and evidence levels.

2004). For those who have more than two major risk factors, a 10-yr risk of symptomatic CHD development should also be calculated as outlined by the National Cholesterol Education Program to determine whether there is a need for primary prevention strategies (I,B) (National Cholesterol Education Program 2004). Patients with a CHD risk equivalent such as abdominal aortic aneurysm, diabetes mellitus, peripheral vascular disease, carotid artery disease, or a 10-yr Framingham cardiac risk of greater than 20%, should be evaluated for appropriate secondary preventative measures as should those with detected CHD (I,A) (Antman et al. 2004).

Physicians must properly educate patients of their STEMI risk level and how to recognize STEMI (I,C). Chest discomfort, weakness, dyspnea, diaphoresis, and nausea are all among the presenting symptoms of STEMI and if there is suspicion of STEMI there should be transport to a hospital via ambulance (I,B) (Antman et al. 2004). Patients should be counseled to call 9-1-1 if the symptoms last longer than 5 min, even in the face of uncertainty regarding the symptoms (I,C). Patients already prescribed nitroglycerin should take one dose of nitroglycerin sublingually upon the onset of chest discomfort, and if discomfort lasts for 5 min, call 9-1-1 (I,C) (Antman et al. 2004). It should be noted that this recommendation replaces an earlier recommendation which entailed three doses of sublingual nitroglycerin each 5 min apart before 9-1-1 was called.

ONSET OF STEMI

Initial response to out-of-hospital cardiac arrest requires community response and, thus, there must be a proper system set in place. In addition to having a system, it is necessary that the community be educated about the system and how to properly implement it. This response system should include recognition of the problem, activation of the emergency medical system (EMS), initiation of cardiopulmonary resuscitation (CPR), defibrillation when appropriate, and access to advanced cardiac life support (ACLS) (I,C). Family members of those at moderate to high risk of having a STEMI need to assume a slightly larger responsibility in the community when it comes to learning CPR and the proper use of automated external defibrillators (AEDs) (I,B) (Antman et al. 2004).

PREHOSPITAL ISSUES

Activation of EMS stretches beyond the patient and those around him/her. Dispatchers responding to 9-1-1 calls must be properly trained with standard protocols to ensure that EMS activation is done in an efficient manner (I,C). While EMS activation is an important factor in responding to STEMI, activation is useless if EMS is not properly prepared to manage the situation. EMS responders must be trained to recognize when defibrillation is necessary and be equipped and trained to provide defibrillation when needed (I,A). Outside of EMS responders, local public safety responders should also be able to manage those with chest pain and suspected cardiac arrest and should be familiar with the use of AEDs (I,B) (Antman et al. 2004).

When 9-1-1 dispatch is contacted, it is appropriate for the dispatcher to advise that the patient chew 162 to 325mg of aspirin if they are not allergic to aspirin while awaiting EMS (IIa,C). If this is not advised to the patient by dispatch, EMS should administer this dose of chewable aspirin to patients (I,C) (Eisenberg and Topol 1996). Initial EMS responders should also obtain and evaluate a 12-lead electrocardiogram (ECG) in patients with suspected STEMI (IIa,B). If

ECG shows findings of STEMI, EMS should contact the destination medical facility and convey pertinent information to them (IIa,C) (Antman et al. 2004).

Trials investigating fibrinolytic therapy have found benefit in beginning therapy as soon as possible (Fibrinolytic therapy trialists' collaborative group 1994). A prehospital fibrinolytic protocol can be established and utilized by EMS if there is a physician in the ambulance or if it is a high volume EMS system in which paramedics are well trained in ECG interpretation, ECG transmission is possible, online medical command is available, and there are constant quality-improvement programs (I,B) (Antman et al. 2004).

Once EMS has arrived and is ready to transport the patient, a proper decision must be made in regards to the destination facility. Those suffering a STEMI with cardiogenic shock and younger than 75 yr old should be taken immediately, or secondarily transported, to a facility that is able to perform cardiac catheterization and percutaneous coronary intervention (PCI) or coronary artery bypass grafting (CABG) within 18 hr of shock onset (I,A). STEMI patients, in whom fibrinolysis is contraindicated, should be taken to, or secondarily transferred to, medical facilities capable of PCI or CABG (I,B). Door-to-departure time should be less than 30 min for both primary transport and secondary transfer. Protocols should be written to help guide EMS providers in deciding where to take STEMI patients in the particular community (I,C). For those suffering a STEMI with cardiogenic shock who are older than 75 yr old, guidelines are the same as those under 75 except the level of evidence is IIa,B (Fig. 5.1) (Antman et al. 2004).

INITIAL RECOGNITION IN THE EMERGENCY DEPARTMENT

Multidisciplinary healthcare teams should be established to develop written protocols for triage and management of patients who come to the emergency department (ED) with possible STEMI (I,B). Guidelines should dictate selection of treatment modality. In situations where ED physicians are unable to make a definitive diagnosis or treatment selection a cardiology consult should be obtained immediately (I,C). No more than 30 min should lapse between a patient's arrival to the ED or contact with EMS and the initiation of fibrinolysis. If PCI is

Management of ST-Elevation Myocardial Infarction 79

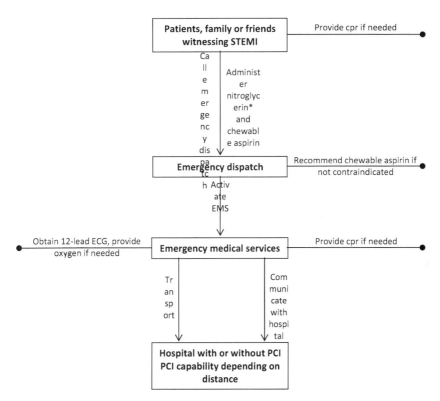

Figure 5.1 What must occur during early response to ST-elevation myocardial infarction.

*Only if prescribed by physician earlier.

Original figure by Loomba et al. STEMI= ST-elevation myocardial infarction, cpr= cardiopulmonary resuscitation, ECG= electocradiogram, PCI= percutaneous coronary intervention, EMS= emergency medical services

the chosen method of management, the door to needle time should not exceed 30 min and door to balloon time should not exceed 90 min (I,B) (Antman et al. 2004).

With time being of the essence in managing patients with suspected STEMI, patient history collected should be focused. History obtained in the ED should include previous occurrences of angina, MI, CABG, or PCI. The character of chest discomfort and any associated symptoms are also vital, as is a complete assessment of comorbidities (I,C) (Antman et al. 2004).

The focused history should be followed by a focused physical exam which focuses on assessment of the extent and localization of STEMI symptoms. The physical should also include a focused neurological examination to assess for previous stroke or cognitive deficits before fibrinolysis (I,C) (Antman et al. 2004).

With a suspected STEMI patient in the ED, a 12-lead ECG should be obtained and read by an ED physician within 10 min of ED arrival (I,C). Serial ECGs at 5 to 10 min intervals, or continuous 12-lead ECG monitoring, should be obtained for patients who remain symptomatic but had a nondiagnostic initial ECG (I,C). Right-sided ECG leads should be obtained to screen for ST elevation and right ventricular infarction in those with inferior STEMI (I,B) (Table 5.2) (Antman et al. 2004).

Table 5.2 Diagnosing right ventricular myocardial infarction with electrocardiogram.

➤ Often found concurrently with inferior wall myocardial infarction noted in leads II, III, aVF	➤ When inferior wall myocardial infarction is observed obtain right sided leads V4R, V5R, V6R	➤ ECG may show ST-segment elevation of ≥1 mm with doming in right sided leads
➤ ST-segment elevation in V4R is usually a sign of right ventricular myocardial infarction but is transient	➤ ST-elevation in lead III > lead II should raise suspicion of right ventricular myocardial infarction	➤ Look for AV block which is commonly seen with right ventricular myocardial infarction

Original table by Loomba et al. 12-lead electrocardiogram findings that can assist in diagnosing right ventricular myocardial infarction.

Serum cardiac biomarkers are also part of making the STEMI diagnosis: while they should be performed, they should not delay reperfusion therapy (I,C). Evaluation of STEMI patients should include cardiac-specific troponins (I,C) with serial troponin levels being obtained to help monitor STEMI patients, particularly for patients receiving only fibrinolytic therapy rather than PCI or CABG within 24 hr of symptom onset (II,B) (Case definitions for acute coronary heart disease in epidemiology and clinical research studies: a statement from the AHA Council on Epidemiology and Prevention; AHA Statistics committee; World Heart Federation Council on Epidemiology and Prevention 2003). In respect to reinfarction, serial troponins or other biomarkers are not reliable for diagnosis with 18 hr of symptom onset (I,C). Point-of-care assays can be used to qualitatively assess cardiac biomarker elevation but should be confirmed with a quantitative test (I,B). Reperfusion therapy, however, should not be delayed for this confirmation and should not be delayed if qualitative assay is negative in the face of STEMI symptoms and ST elevation on 12-lead ECG (I,C) (Antman et al. 2004).

Portable chest x-rays should be obtained for STEMI patients, but as with cardiac biomarkers, imaging should not delay reperfusion therapy unless aortic dissection is suspected. In the instance that aortic dissection is suspected and diagnosis is unclear, imaging such as portable x-ray, echocardiography, CT scan, or MRI should be used to make a definitive diagnosis (I,B). Single photon emission computed tomography imaging does not have a role in diagnosing STEMI in those who already have ECG findings confirming the diagnosis (IIa,B) (Antman et al. 2004).

INITIAL MANAGEMENT IN THE EMERGEMCY DEPARTMENT

Certain management measures should be taken as soon as possible when STEMI is suspected. Patients with an arterial oxygen saturation of less than 90% should receive supplemental oxygen (I,B). It is not inappropriate, however, to give all STEMI patients supplemental oxygen within 6 hr of symptom onset (I,C) (Antman et al. 2004).

Ongoing chest discomfort should be countered with 0.4 mg sublingual nitroglycerin every 5 min with a maximum of three doses, at which point the need for intravenous nitroglycerin should be assessed (I,C). Intravenous nitroglycerin is appropriate to use to help relieve chest discomfort, control hypertension, and manage pulmonary congestion (I,C). Patients should not receive nitrates if systolic blood pressure is below 90 mm Hg or 30 mm Hg below baseline, heart rate is less than 50 beats per min or greater than 100 beats per min, or there is suspicion of right ventricle infarction (I,C). Those having taken a phosphodiesterase inhibitor for erectile dysfunction in the last 24–48 hr should not receive nitrates either (I,B) (Antman et al. 2004).

Pain management can be accomplished with 2 to 4 mg of intravenous morphine sulfate at 2 to 8 mg increments at intervals of 5 to 15 min (I,C) (Antman et al. 2004). Patients taking NSAIDs, other than aspirin, before STEMI should have these discontinued due to an increased risk of reinfarction, hypertension, heart failure, myocardial rupture, and mortality (I,C) (Yusuf et al. 2006, Antman et al. 2008).

While aspirin should be administered to the patient upon arrival of EMS or even before that, if a patient has not received aspirin since symptom onset, the ED should give 162 (I,A) to 325 mg (I,C) of chewable aspirin. This can be continued indefinitely with a daily dose of 75 to 162 mg (Antithrombotic Trialists' Collaboration 2002). Enteric-coated aspirin can be used but non-enteric-coated aspirin undergoes more rapid buccal absorption (Antman et al. 2004).

Beta-blockers are an important aspect in the initial management of STEMI. Those without contraindications to beta-blockers should receive intravenous beta-blockers regardless of current or future fibrinolytic therapy or PCI (I,B), particularly if tachyarrhythmias or hypertension are also a concurrent issue (I,C). Prompt beta-blocker administration in STEMI patients helps reduce the extent of infarction, incidence of reinfarction, and incidence of ventricular arrhythmias (Antman et al. 2004).

While the patient is initially being triaged and managed it is important for reperfusion evaluation to occur as well so that a reperfusion plan can be established and executed quickly (I,A). Whether reperfusion is planned with fibrinolysis or PCI, maintaining a short time to reperfusion is vital as both short and long term outcomes are determined, largely, by this time interval (De luca et al. 2003). Selection of reperfusion strategy is highly variable from patient to patient but should be based upon factors such as time since onset of symptoms, risk of STEMI mortality, risk of bleeding, and time to nearest skilled PCI laboratory (Antman et al. 2004).

FIBRINOLYTIC THERAPY

For STEMI patients at a medical center without the ability for skilled PCI within 90 min of initial medical contact, fibrinolysis should be the chosen reperfusion method if there are no contraindications (I,B) (Antman et al. 2008). Fibrinolysis should be administered within

12 hr of symptom onset and ST elevation greater than 0.1 mV in at least two consecutive precordial or one adjacent limb leads, or in those with new left bundle branch block (I,A) (Fibrinolytic Therapy Trialists' Collaborative Group 1994). It is also appropriate to start fibrinolytic therapy within 12 hr of symptom onset if 12-lead ECG findings are indicative of a posterior MI (IIa,C). For patients in whom symptom onset began over 24 hr earlier (III,C) or 12-lead ECG shows ST-segment depression without suspicion of posterior MI, fibrinolytic therapy should not be started (III,A) (Antman et al. 2004).

Fibrinolytic therapy may be combined with glycoprotein IIb/IIIa inhibitors such as abciximab and half-dose reteplase or tenecteplase in order to prevent reinfarction (IIb,A). Studies have shown that this reinfarction prevention does not necessarily mean higher survival at either 30 d or a yr (IIb,B) (Lincoff et al. 2002). This combination therapy should not be used in patients older than 75 yr old or history of ICH due to increased risk of ICH (III,B) (Antman et al. 2004).

The following contraindications must be kept in mind when considering fibrinolytic therapy: history of intracranial hemorrhage (ICH), closed head or facial trauma within past 3 mon, ischemic stroke within past 3 mon, and uncontrolled hypertension (I,A). Complications of fibrinolytic therapy include change in neurological status during or after reperfusion therapy, which if within 24 hr of treatment initiation, must be managed as ICH until proven otherwise. Fibrinolytic, antiplatelet, and anticoagulant agents must all be discontinued at this point until imaging shows no evidence of ICH (I,A). When ICH occurs, a neurology or hematology consult be obtained and appropriate infusions of cryoprecipitate, fresh frozen plasma, protamine, and platelets should be administered when needed (I,C) (Antman et al. 2004).

PERCUTANEOUS CORONARY INTERVENTION

Primary PCI should be considered if immediately available for STEMI patients within 12 hr of symptom onset and a door-to-balloon time of 90 min can be achieved by those skilled in the procedure (I,A). If symptom duration is less than 3 hr and needle to balloon time is less than 1 hr, then primary PCI should be performed over fibrinolytic therapy (I,B). Patients younger than 75 yr old with ST elevation who develop shock within 36 hr of MI and are candidates for revascularization within 18 hr of shock should also get PCI (I,A). Those with severe congestive heart failure or pulmonary edema and onset of symptoms within 12 hr are also candidates for PCI (I,B). PCI is also appropriate for some patients who are over 75 yr old with STEMI if they develop shock within 36 hr of STEMI and can undergo revascularization within 18 hr of shock (I,B) (Antman et al. 2004).

If PCI for STEMI is performed by physicians performing less than 75 PCI procedures a year, benefit over fibrinolysis is not clear (IIb,C). PCI of a noninfarct artery should not be performed during primary PCI in patients that are hemodynamically stable and PCI should not be undertaken in asymptomatic patients after 12 hr of STEMI onset if the patient is hemodynamically and electrically stable (III,C) (Antman et al. 2004). PCI of a totally occluded artery beyond 24 hr of STEMI onset is not advisable in patients who are asymptomatic and have one or two vessel disease (III,B) (Antman et al. 2008).

For STEMI patients who are ineligible for fibrinolysis, primary PCI should be undertaken within 12 hr of symptom onset (I,C). STEMI patients ineligible for fibrinolysis who have severe CHF, hemodynamic or electrical instability, or persistent ischemic symptoms can undergo PCI 12 to 24 hr after symptom onset (IIa,C) (Antman et al. 2004).

Studies have compared primary stenting and primary angioplasties and have found no significant difference between the two in mortality or reinfarction. The rate of major adverse cardiac events was reduced in those with stents, however (Scheller et al. 2001, Zhu et al. 2001). In respect to bare metal stents versus drug eluting stents, there was no significant difference in mortality, reinfarction, or need for revascularization at 30 d (Lemos et al. 2004).

STEMI patients who undergo PCI experience less short-term mortality, less nonfatal reinfarction, and less hemorrhagic stroke when compared with those who undergo fibrinolysis. Those undergoing PCI, however, do show an increase in major bleeding risk (Keeley et al. 2003). It appears that the greatest mortality benefit with PCI is in high-risk patients (Antman et al. 2004).

Nonprimary PCI may be performed in high-risk patients if PCI is not available initially and bleeding risk from earlier treatment or comorbidities is low (I,C) (Antman et al. 2008).This should be done only after fibrinolytic regiments that are not full-dose (III,B) (Antman et al. 2008, Keeley et al. 2006, Montalescot et al. 2007). PCI after fibrinolysis may also be done in patients who have evidence of reinfarction after fibrinolysis, moderate to severe spontaneous provocable myocardial ischemia after STEMI (I,B), cardiogenic shock, or hemodynamic instability (I,C) (Antman et al. 2004).

ACUTE SURGICAL REPERFUSION

CABG should be performed in STEMI patients if PCI has failed and there is remaining discomfort, if PCI has failed and there is hemodynamic instability, recurrent ischemia refractory to medical therapy in those who are not candidates for PCI of fibrinolysis (I,B), in those less than 75 yr old with cardiogenic shock with STEMI and have severe multivessel or left main disease. Patients with stable angina and 70% left main coronary artery stenosis, those with three vessel disease, those with two-vessel disease with 70% proximal left anterior descending coronary artery stenosis with ejection fraction less than 0.50 are all candidates for CABG (I,A) (Antman et al. 2004).

Emergency CABG should not be undertaken in those who are hemodynamically stable with a small myocardial area at risk despite persistent angina or in those with successful epicardial reperfusion but unsuccessful microvascular reperfusion (III,C) (Antman et al. 2004).

Mortality associated with CABG is higher within the first 3 to 7 d after STEMI and so this risk must be balanced with any potential benefit when selecting reperfusion strategies. If CABG is necessary in patients who no longer have ongoing ischemia and are otherwise hemodynamically stable, surgery should be delayed for several days to allow for myocardial recovery (I,B) (Antman et al. 2004).

Patients undergoing elective CABG who are taking clopidogrel should stop the drug for 5 to 7 d before surgery. Those taking aspirin do not have to stop taking the drug before CABG (I,B) (Antman et al. 2004, Yusuf et al. 2001).

ANCILLARY ANTICOAGULANT THERAPY

Various ancillary therapies have been studied with the reperfusion strategies available. Patients undergoing PCI or CABG should receive unfractionated heparin(UFH)(I,C). Those undergoing reperfusion with alteplase, reteplase, or tenecteplase should receive a

60 U/kg bolus (4000 U maximum) of UFH followed by an infusion of 12 U/kg per hr (1000 U/hr maximum) with a PTT being monitored and maintained at 1.5 to 2.0 times normal (I,C). For patients being treated with streptokinase, anistreplase, or urokinase and at high risk for systemic embolization, intravenous UFH should be given as well (I,B). Intravenous UFH may be administered to those receiving streptokinase as well (IIb,B). For all patients receiving UFH, platelet counts should be monitored daily (I,C) (Antman et al. 2004). Patients undergoing fibrinolysis should receive at least 48 hr of anticoagulation (I,C). Ideally, anticoagulation should last for the entire hospitalization up to a maximum of 8 d with an anticoagulant other than UFH being used if anticoagulation lasts over 48 hr to avoid heparin induced thrombocytopenia (I,A) (Antman et al. 2008).

Low molecular weight heparin (LMWH) can be used as an alternative to UFH for patients less than 75 who are undergoing fibrinolysis and donot have significant renal dysfunction. The combination of Enoxaparin and tenecteplase has been used for many studies with a 30 mg IV bolus of enoxaparin followed by a 1.0 mg/kg subcutaneous injection every 12 hr until discharge (IIb,B). LMWH should be avoided in patients over 75 yr old undergoing fibrinolysis and in those under 75 yr old with significant renal dysfunction (III,B) (Antman et al. 2004).

Patients for whom heparin may not be an option due to previously documented heparin induced thrombocytopenia (HIT), bivalirudin is an alternative anticoagulant to be used with streptokinase and can be administered with an initial bolus of 0.25 mg/kg followed by an intravenous infused of 0.5 mg/kg per hour for 12 hr and then 0.25 mg/kg per hr for another 36 hr (IIa,B) (Antman et al. 2004, Thrombin specific anticoagulation with bivalirudin versus heparin in patients receiving fibrinolytic therapy for acute myocardial infarction: the HERO-2 randomized trial 2001).

When managing ancillary anticoagulation regimens in patients who will be undergoing PCI, there are recommendations to keep in mind. For patients in whom UFH was administered, additional boluses can be given as needed for the procedure, bearing in mind any GPIIb/IIIa inhibitors that may have been given (I,C). For enoxaparin, if the last subcutaneous dose was within 8 hr of PCI then no additional doses should be given. If the last dose was 8–12 hr before PCI, however, an IV dose of 0.3 mg/kg should be given

(I,B). Fondaparinux should not be used as ancillary anticoagulation to support PCI due to the increased risk of catheter thrombosis (III,C) (Antman et al. 2008).

STEMI patients not receiving reperfusion therapy may also receive anticoagulation therapy for the duration of the hospitalization up to a maximum of 8 d, although UFH should not be used in these particular patients (I,B) (Antman et al. 2008).

ANCILLARY ANTIPLATELET THERAPY

All STEMI patients without an aspirin allergy should take daily oral aspirin with an initial dose of 162 mg to 325 mg and a maintenance dose of 75 mg to 162 mg for an indefinite period of time (I,A) (Antman et al. 2004). A 75 mg daily dose of oral clopidogrel should be given in addition to aspirin in STEMI patients regardless of reperfusion plans (I,A) for at least 14 d (I,B) (Antman et al. 2008).

Those who will be undergoing primary PCI should receive a 300 to 600 mg loading dose of clopidogrel as soon as possible before primary or non primary PCI (I,C) or 60 mg prasugrel (I,B) as soon as possible before primary PCI. Patients who are on clopidogrel (I,B) or prasugrel (I,C) and need CABG should withhold clopidogrel for 5 to 7 d prior to the surgery unless the surgery is urgently needed and the benefits of surgery are believed to outweigh the bleeding risks (Kushner et al. 2009). For patients undergoing fibrinolysis in whom aspirin is contraindicated, clopidogrel can be used as the ancillary antiplatelet of choice (IIa,C) (Antman et al. 2004). A 300 mg oral clopidogrel loading dose is also appropriate in patients under 75 yr old who receive fibrinolytic therapy or no reperfusion therapy (IIa,C) (Antman et al. 2008).

Combination antiplatelet therapy utilizing clopidogrel and aspirin is recommended for STEMI patients who undergo stent implantation (Antman et al. 2004, Mehta et al. 2001, Steinbuhl et al. 2002). Prasugrel, however, should not be used in any dual antiplatelet regimen in patients with a history of stroke or transient ischemic attack (Kushner et al. 2009).

Glycoprotein IIb/IIIa inhibitors can also be used as ancillary antiplatelet therapy. Abciximab can be started at the time of primary PCI in STEMI patients (IIa,A) as can tirofiban or eptifibatide (IIa,B)

(Lee et al. 2003). The utility of glycoprotein IIb/IIIa inhibitors in preparation for PCI is uncertain at this time (IIb,B) (Antman et al. 2004, Antman et al. 2008).

HOSPITAL MANAGEMENT

In the coronary care unit (CCU), care must be closely based on appropriately derived protocols (I,C). STEMI patients should be in a quiet, comfortable environment where continuous ECG and pulse oximetry monitoring can take place along with easy access to hemodynamic monitoring and defibrillation (I,C). Medications should be closely reviewed to ensure adequate aspirin and beta-blocker dosing for heart rate control along with an assessment of the need for IV nitroglycerin for control of chest discomfort and hypertension (I,A). Arterial oxygen saturation also needs to be monitored with supplemental oxygen appropriate if saturation drops below 90% (I,C). Terminally ill patients, who wish to not be resuscitated, should not be admitted to the CCU (Antman et al. 2004).

Low-risk STEMI patients can be admitted directly to a stepdown unit after PCI rather than the CCU (I,C). Those admitted to the CCU can be transferred to the stepdown unit after 12 to 24 hr of clinical stability (I,C). STEMI patients with clinically symptomatic heart failure or arrhythmias but are hemodynamically stable can also be managed in the stepdown unit (I,C) (Antman et al. 2004).

INPATIENT MEDICATION ASSESSMENT

While STEMI patients are recovering in the hospital it is important to adequately assess the medications they are on routinely.

Beta-blockers are an important aspect of the medical regimen as a significant risk reduction in both short and long-term mortality has been found in those started on beta-blockers after STEMI (Chae and Hennekens 1999). Beta-blocker therapy initiated within the first 24 hr of STEMI should be continued during the early convalescent stage (I,A). If beta-blockers are not started within the first 24 hr of STEMI and there are no contraindications, beta-blockers should be started in the early convalescent stage (I,A). For patients with contraindications to beta-blocker therapy in the first 24 hr of STEMI, a re-evaluation is appropriate to gauge whether beta-blocker therapy can be started later (I,C) (Antman et al. 2004).

IV Nitroglycerin should be used in the first 48 hr if there is persistent chest discomfort, hypertension, or CHF but its use should not hinder the administration of mortality reducing treatment elements such as beta-blockers or ACE inhibitors (I,B). After the first 48 hr, IV, oral, or topical nitrates can also be used for continuing chest discomfort or CHF as long nitrate use does not hinder the use of beta-blockers and ACE inhibitors (I,B). Use of nitrates outside the initial 48 hr may have beneficial effects even if persistent chest discomfort or CHF are not present, although this benefit may not be large and is not well understood (IIb,B). Patients with a systolic blood pressure of less than 90 mm Hg and greater than 30 mm Hg, heart rate of greater than 100 beats per min, or heart rate less than 50 beats per min should not be given nitrates (III,C) (Antman et al. 2004).

ACE inhibitors are another important piece of the medical regimen in STEMI patients. An oral ACE inhibitor should be given during the convalescent stage and continued long-term in those who can tolerate them (I,A). Those who cannot tolerate ACE inhibitors can be started on an ARB, such as valsartan or candesartan, if there are signs of heart failure or ejection fraction of less than 0.40 (IIa,B) (Dickstein and Kjekshus 2002, Pfeffer et al. 2003). Studies have shown that ACE inhibitors offer the greatest benefit in higher-risk patients such as those with history of MI, CHF, or low ejection fraction (Antman et al. 2004, Latini et al. 1995, ACE Inhibitor Myocardial Infarction Collaborative Group 1998, Flather et al. 2000).

Aldosterone blockade should also be prescribed for patients after STEMI if they already receiving an ACE inhibitor, have an ejection fraction of less than 0.40, or symptomatic heart failure (I,A). Contraindications include hyperkalemia and renal dysfunction (III,A) (Antman et al. 2004).

Oxygen, antiplatelet therapy, and antithrombotic therapy should also be maintained as outlined previously.

CONCLUSION

It is important for healthcare providers to be aware of the STEMI guidelines and keep up to date on any changes. Strategies should be implemented to make sure that guidelines are followed, ensuring optimal STEMI management and reduction in associated morbidity and mortality.

KEY FACTS

- STEMI occurs when an artery supplying the heart gets fully blocked.
- STEMI is one of the leading causes of in-hospital death.
- The response to STEMI involves the community and physicians, among others.
- Management guidelines have been developed to improve care of STEMI.
- Time to opening the blocked artery is vital to care and amount of heart muscle that can be saved.
- Strategies are needed to make sure management guidelines are implemented properly.

SUMMARY

- ST-Elevation myocardial infarction is one of the leading causes of in-hospital mortality and following evidence based guidelines is important to ensure the best management and reducing mortality rate.
- ST-elevation myocardial infarction response begins with primary prevention and patient education.
- Response to ST-elevation myocardial infarction involves others than the patient and physicians as community bystanders and emergency medical services also play a large role in response
- Emergency medical services must appropriately pick a healthcare facility for ST-elevation myocardial infarction patients based on the distance and the presence of a cath lab.
- Emergency department care must focus on quickly establishing a diagnosis via electrocardiogram and cardiac enzymes, providing relief with oxygen and nitrates, and beginning the decision of revascularization method.

- Revascularization, whether it be via fibrinolysis, percutaneous coronary intervention, or coronary artery bypass grafting, must begin as soon as possible.
- In-hospital management must include appropriate pharmacologic therapy.

DEFINITIONS

Angina: Chest pain or discomfort caused by a partial or complete block of an artery supplying the heart.

Anticoagulation: The pharmacologic process of lowering the blood's ability to clot to prevent blood clots.

Antiplatelet therapy: The pharmacologic process of lowering the ability of platelets to aggregate and form clots.

Coronary artery: An artery supplying oxygenated blood to the heart.

Coronary artery bypass grafting: The surgical process of bypassing a blocked portion of an artery by using a piece of another vessel.

Fibrinolysis: The pharmacologic process of destroying blood clots already formed and in the circulation.

Ischemia: The process of tissue death which can occur after the loss of blood supply as occurs from blockage of an artery.

Percutaneous coronary intervention: Catheter based approach to image the coronary arteries which also allows for catheter based procedures to deal with blockages.

Revascularization: The process of eliminating a blockage in an artery.

ST-Elevation myocardial infarction: Heart muscle damage caused when blood flow to the heart muscle is restricted by a blockage in one of the arteries supplying the heart.

LIST OF ABBREVIATIONS

ACC	:	American College of Cardiology
ACE	:	Angiotensin converting enzyme
ACLS	:	Advanced cardiac life support

AED	:	Automated external defibrillator
AHA	:	American Heart Association
CABG	:	Coronary artery bypass graft
CCU	:	Coronary care unit
CHD	:	Coronary heart disease
CPR	:	Cardiopulmonary resuscitation
ECG	:	Electrocardiogram
ED	:	Emergency department
EMS	:	Emergency medical system
HIT	:	Heparin induced thrombocytopenia
ICH	:	Intracranial hemorrhage
LMWH	:	Low molecular weight heparin
PCI	:	Percutaneous coronary intervention
STEMI	:	ST elevation myocardial infarction
UFH	:	Unfractionated heparin

REFERENCES CITED

ACE Inhibitor Myocardial Infarction Collaborative Group. 1998. Indications for ace inhibitors in the early treatment of acute myocardial infarction: systematic overview of individual data from 100,000 patients in randomized trials. Circulation 97: 2202–12.

Antithrombotic Trialists' Collaboration. 2002. Collaborative meta-analysis of randomised trials of antiplatelet therapy for prevention of death, myocardial infarctin, and stroke in high risk patients. BMJ 324: 71–86.

Antman, E.M., D.T. Anbe, P.W. Armstrong, E.R. Bates, L.A. Green and M. Hand. 2004. ACC/AHA guidelines for the management of patients with ST-elevation myocardial infarction-executive summary: a report of the American College of Cardiology/American Heart Association Task Force on Practice Guidelines. Circulation 110: 588–636.

Antman, E.M., M. Hand and P.W. Armstrong. 2008. 2007 focused update of the ACC/AHA 2004 guidelines for the management of patients wit ST-elevation myocardial infarction. Circulation 117: 296–329.

Chae, C., C.H. Hennekens. 1999. Clinical trials in cardiovascular disease: a companion to braunwald's heart disease. WB Saunders. Philadelphia, PA, USA.

Chen, J., M.J. Radford and Y. Wang. 2000. Are beta blockers effective in elderly patients who undergo coronary revascularization after acute myocardial infarction. Arch Intern Med 160: 947–52.

De luca, G., H. Suryapranata and F. Zijlstra. 2003. Symptom-onset-to-balloon time and mortality in patients with acute myocardial infarction treated by primary angioplasty. J Am Coll Cardiol 42: 991–97.

Dickstein, K. and J. Kjekshus. 2002. Effects of losartan and captopril on mortality and morbidity in high-risk patients after acute myocardial infarction: the OPTIMAAL randomised trial: optimal trial in myocardial infarction with antiotensin II antagonish losartan. Lancet 360: 752–60.

Eisenberg, M.J. and EJ Topol. 1996. Prehospital administration of aspirin in patients with unstable angina and acute myocardial infarction. Arch Intern Med 156: 1506–10.

Fibrinolytic Therapy Trialists' Collaborative Group. 1994. Indications for fibrinolytic therapy in suspected acute myocardial infarction: collaborative overview of early mortality and morbidity results from all randomized trials of more than 1000 patients. Lancet 343: 311–22.

Flather, M.D., S. Yusuf and L. Kober. 2000. Long-term ACE inhibitor therapy in patients with heart failure or left ventricular dysfunction: a systematic overview of data from individual patients. Lancet 355: 1575–1591.

Gutstein, D.E. and V. Fuster. 1998. Pathophysiologic bases for adjunctive therapies in the treatment and secondary prevention of acute myocardial infarction. Clin Cardiol 21: 161–8.

Keeley, E.C., J.A. Boura and C.L. Grines. 2003. Primary angioplasty versus intravenous thrombolytic therapy for acute myocardial infarction: a quantitative review of 23 randomised trials. Lancet 361: 13–20.

Keeley, E.C., J.A. Boura and C.L. Grines. 2006. Comparison of primary and facilitated percutaneous coronary interventions for ST-elevation myocardial infarction: quantitative review of randomised trials. Lancet 367: 579–88.

Kushner, F.G., M. Hand and S.C. Smith. 2009. 2009 Focused Updates: ACC/AHA guidelines for the management of patients with ST-elevation myocardial infarction. J Am Coll Cardiol 54: 2205–41.

Latini, R., A.P. Maggioni and M. Flather. 1995. ACE inhibitor use in patients with myocardial infarction: summary of evidence from clinical trials. Circulation 92: 3132–37.

Lee, D.P., N.A. Herity and B.L. Hiatt. 2003. Adjunctive platelet glycoprotein IIb/IIIa receptor inhibition with tirofiban before primary angioplasty improves angiographic otcomes: results of the tirofiban given in the emergency room before primary angioplasty (TIGER-PA) pilot trial. Circulation 107: 1497–01.

Lemos, P.A., F. Saia and S.H. Hofma. 2004. Short and long term clinical benefit of sirolimus eluting stents compared to conventional bare stents for patients with acute myocardial infarction. J Am Coll Cardiol 43: 704–08.

Lincoff, A.M., R.M. Califf and F. van de Werf. 2002. Mortality at 1 year with combination platelet glycoprotein IIb/IIIa inhibition and reduced dose fibrinolytic therapy vs conventional fibrinolytic therapy for acute myocardial infarction. JAMA 288: 2130–35.

Mehta, S.R., S. Yusuf and R.J. Peters. 2001. Effects of pretreatment with clopidogrel and aspirin followed by long term therapy in patients undergoing percutaneous coronary intervention: the PCI-CURE study. Lancet 358: 527–33.

Montalescot, G., D. Antoniucci and A. Kastrati. 2007. Abciximab in primary coronary stenting of ST-elevation myocardial infarction: a European meta-analysis on individual patients' data with long term follow up. Eur Heart J 28: 443–9.

National Cholesterol Education Program. 2004. Third report of the expert panel on detection, evaluation, and treatment of high blood cholesterol in adults (Adult treatment panel III). National Heart Lung and Blood Institute. National Institutes of Health. Retrieved December 30, 2010 (http://www.nhlbi.nih.gov/guidelines/cholesterol/index.htm).

Pfeffer, M.A., J.J. Mcmurray and E.J. Velazquez. 2003. Valsartan, captopril, or both in myocardial infarction complicated by heart failure, left ventricular dysfunction, or both. N Engl J Med 349: 1893–06.

Scheller, B., B. Hennen and S. Severin-kneib. 2001. Long term follow up of a randomized study of primary stenting versus angioplasty in acute myocardial infarction. Am J Med 110: 1–6.

Steinbuhl, S.R., P.B. Berger and J.T. Mann. 2002. Early and sustained dual oral antiplatelet therapy following percutaneous coronary intervention: a randomized controlled trial. JAMA 288: 2411–20.

Wiviott, S.D., D.A. Morrow and R.P. Giugliano. 2003. Performance of the thrombolysis in myocardial infarction risk index for early acute coronary syndrome in the National Registry of Myocardial Infarction: a simple risk index predicts mortality in both ST and non-ST elevation myocardial infarction. J Am Coll Cardiol 41: 365A–366A.

Yusuf, S., F. Zhao and S.R. Mehta. 2001. Effects of clopidogrel in addition to aspirin in patients with acute coronary syndromes without ST-segment elevation. N Engl J Med 345: 494–02.

Yusuf, S., S. Ounpuu, T. Dans, A. Avezum, F. lanas and M. Mcqueen. 2004. Effect of potenitally modifiable risk factors associated with myocardial infarction in 52 countries (the INTERHEART study): case-control study. Lancet 364: 937–52.

Yusuf, S., S.R. Mehta and S. Chrolavicius. 2006. Effects of fondaparinux on mortality and reinfarction in patients with acute ST-segment elevation myocardial infarction: the OASIS-6 trial. JAMA 295: 1519–30.

Zhu, M.M., A. Feit and H. Chadow. 2001. Primary stent implantation compared with primary balloon angioplasty for acute myocardial infarction: a meta analysis of randomized clinical trials. Am J Cardiol 88: 297–301.

6

Non-Invasive Coronary Angiography with Cardiac CT in Patients with Angina Pectoris

Christof Burgstahler,[1], Harald Brodoefel[2] and Stephen Schroeder[3]*

ABSTRACT

Multi-slice computed tomography has become an important diagnostic tool in the care of patients with cardiovascular disease. Since 1999 a rapid technical evolution from 4-slice scanners to 128-slice and— meanwhile 320-slice—scanners with faster tube rotation has taken place. Due to these advances as well as more dedicated post-processing tools cardiac CT allows assessing cardiac structures with increasing spatial and temporal resolution. Today cardiac computed tomography is not only able to image the coronary arteries but to get information about cardiac function and structure. With up-to-date scanners reliable information about coronary arteries, left and right ventricular function, myocardial viability or valvular heart disease can be achieved within

---

[1]Department of Internal Medicine V–Sports Medicine, University of Tuebingen, Silcherstrasse 5, D–72076 Tübingen;
Email: christof.burgstahler@med.uni-tuebingen.de; Burgstahler@gmx.de
[2]Department of Diagnostic Radiology Eberhard-Karls-University, Tuebingen Germany; Email: h.brodoefel@t-online.de
[3]Klinikum am Eichert, Eichertstrasse 3, D-73035 Goeppingen, Germany;
Email: stephen.schroeder@kae.de
*Corresponding author

List of abbreviations after the text.

one single scan. Moreover, sophisticated acquisition techniques have been employed to reduce radiation exposure to a minimum.

However, the main focus of cardiac CT still remains to visualize the coronary arteries non-invasively. At present there is no non-invasive method that is able to image coronary arteries with a comparable image quality. Especially in patients with low- to intermediate pre-test probability of having significant coronary artery stenosis, cardiac CT is a very helpful work-up tool and might be understood as a 'gate-keeper' for invasive angiography. This chapter provides an overview of non-invasive coronary angiography in patients with angina pectoris and additionally mentions present clinical indications of cardiac computed tomography in clinical cardiology.

INTRODUCTION

Non-invasive imaging with multi-slice computed tomography has become an important diagnostic tool in cardiology practice within the last decade. The rapid technical evolution from 4-slice scanners to 128-slice or 320-slice—scanners with faster tube rotation times as well as improved post-processing tools has led to a stabilization of image quality. This allows the assessment of cardiac structures with high spatial and temporal resolution. At the same time, acquisition techniques have been modified to markedly reduce radiation exposure.

Nowadays cardiac computed tomography has the potential not only to evaluate the coronary arteries, but to get reliable information about cardiac function and cardiac structure.

PRINCIPAL OF CARDIAC COMPUTED TOMOGRAPHY

The key to robust non-invasive cardiac imaging with computed tomography is a rotating X-ray tube with a minimum of 64 slices. The whole heart is covered within one single breath hold after administration of iodinated contrast media. Simultaneous acquisition of an ECG permits 'freezing' the beating heart in different heart phases. Visualization of the coronary arteries is performed by post-processing the acquired dataset, normally in the mid-diastolic phase of the heart cycle. In contrast to the first scanner types with 4-slices, up-to-date scanners permit detection of coronary lesions within the entire coronary tree (Brodoefel et al. 2008a).

Besides the assessment of the coronaries, additional information on cardiac anatomy and morphology such as myocardial bridging, global left and right ventricular function, valve morphology and function, pulmonary veins, myocardial perfusion and viability can be obtained (Schroeder et al. 2008).

At present, MSCT scanners with a minimum of 64 slices and with dedicated algorithms for post-processing are recommended for cardiac computed tomography.

To further improve and stabilize image quality and consequently the diagnostic accuracy, two different philosophies have come into focus within the last few years. Increasing the number of detector rows led to the introduction of volumetric CT scanning (256- or 320-detector row scanners) allowing to scan the whole heart within a single heart beat (Voros 2009), whereas dual source technology is characterized by a heart-rate independent temporal resolution of approximately 83 ms (Achenbach et al. 2006) and the potential of tissue characterization.

PRACTICE AND PROCEDURES—CARDIAC COMPUTED TOMOGRAPHY IN CLINICAL SETTINGS

Coronary Angiography for the Detection of Coronary Stenoses

The growing interest in cardiac CT is mostly based on the ability to visualize coronary arteries non-invasively.

By the use of 4-slice MSCT scanners in the early new millennium, only the proximal parts of the coronary arteries were accessible, whereas today the visualization and assessment of the entire coronary tree with diagnostic image quality has become a reality. Results of comparative studies with invasive coronary angiography are summarized in Table 6.1. These studies indicate a high diagnostic accuracy in the detection of coronary lesions and particularly demonstrate an excellent negative predictive value. That means that cardiac CT can be used to exclude significant coronary lesions (Fig. 6.1). However, most studies have been performed in selected patient populations with an exclusion of patients with higher heart

Table 6.1 Diagnostic accuracy of cardiac computed tomography to detect coronary stenoses in comparison to invasive angiography.

Author	Excluded Pat (n)	Year of Publication	Pat (n)	Excluded Segments %	Sensitivity %	Specificity %	PPV %	NPV %
Ropers(Ropers et al. 2006)	3	2006	81	4 (45/1128)	93	97	56	100
Ong(Ong et al. 2006)	2	2006	134	10 (143/1474)	82	96	83	96
Nikolau(Nikolaou et al. 2006)	4	2006	72	10 (10/1015)	82	95	72	97
Ropers(Ropers et al. 2007)	0	2007	100	0/1394	92	97	68	99
Brodoefel(Brodoefel et al. 2008a)	0	2008	100	0/1300	91	92	75	98
Nasis (Nasis et al. 2010)	0	2010	63	0/973	87	97	73	99

This table summarizes recent publications dealing with the diagnostic accuracy of coronary angiography with cardiac computed tomography. Pat: patients, PPV: positive predicitve value, NPV: negative predictive value. Unpublished table.

rates, advanced coronary heart disease or with irregular heart beats (e.g., atrial fibrillation).

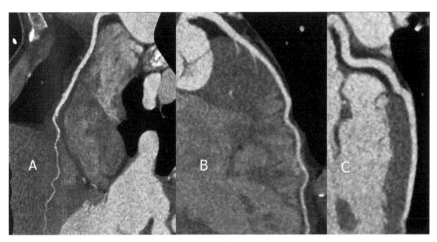

Figure 6.1 Non-invasive coronary angiography with computed tomography. Unpublished images.

These images show a coronary angiography with dual-source computed tomography in a patient with atypical chest pain. All coronaries are free of stenosis. The radiation exposure was approximately 2 mSv. A: Right coronary artery; B: left anterior descending artery; C: left circumflex artery.

On the other hand there is evidence that even patients with suspicion of acute coronary syndromes and normal or equivocal ECG or cardiac biomarkers might be evaluated by cardiac CT.

Cardiac CT is helpful to exclude or diagnose coronary artery disease in selected patient populations.

Some of the initial limitations could be overcome by dual source computed tomography (DSCT). As mentioned above, DSCT is a scanner system with two X-ray tubes mounted in a 90° geometry that allows reducing the temporal resolution to less than approximately 83 ms. With DSCT, up to 98% of all coronary segments can be visualized without motion artifacts, even without lowering the heart rate by administering beta blockers (Achenbach et al. 2006). Latest results indicate that DSCT allows to detect and exclude coronary stenoses also in unselected patients with a high prevalence of CAD and a relevant number of patients without stable sinus rhythm with high accuracy (Brodoefel et al. 2008a). However, the overestimation of stenosis, mostly in calcified segments seems to be a consistent limitation of the method.

Coronary angiography in patients with heart rhythm disorders

In patients with heart rate irregularities, data are controversial. Although image quality and the number of evaluable coronary segments is higher in patients scanned with DSCT, Tsiflikas et al. report on a relatively low sensitivity and positive predictive value (PPV) to detect significant coronary artery disease (sensitivity 73%, PPV 63%) in a cohort of 44 patients without stable sinus rhythm (Tsiflikas et al. 2009). In contrast to these data, Oncel et al. describe a high diagnostic accuracy of DSCT in patients with atrial fibrillation (Oncel et al. 2007). However, up to now, coronary angiography with CT is not recommended in patients with heart rhythms disorders (Schroeder et al. 2008, Taylor et al. 2010).

Imaging of Coronary Artery Bypass Grafts

Thanks to the vessel diameter and course, coronary artery bypass imaging with cardiac CT is characterized by a very high diagnostic accuracy (Fig. 6.2). According to a analysis published in 2008, pooled data estimates for bypass graft analysis were 99% sensitivity, 96% specificity, with a median PPV and NPV values across studies of 93% (range 90 to 95%) and 99% (range 98 to 100%), respectively (Mowatt et al. 2008). Nevertheless, native coronary arteries or the distal run-off is often difficult to evaluate which limits the use of CT in these patients.

Although cardiac CT allows reliable information about bypass grafts, the use of CT is limited as native coronary arteries are mostly heavily calcified in these patients leading to poor image quality in many cases.

MSCT Contrast Enhanced Coronary Angiography for the Assessment of Stents

The visualization of coronary artery stents remains challenging due to metal artifacts caused by the stent struts. Moreover, stent material, stent design and stent diameter as well as CT scanner type are important variables (Schroeder et al. 2008). Based on the present data, the routine application of cardiac CT to assess coronary stents

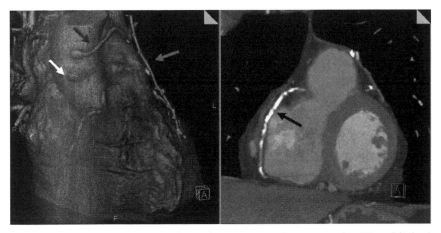

Figure 6.2 Bypass imaging with cardiac computed tomography. Unpublished images.

The left internal mammarian artery is connected to left anterior descending artery (grey arrow). The black arrow indicates a patent venous bypass graft to the circumflex artery whereas another venous bypass graft is occluded (white arrow). The heavily calcified right coronary artery is displayed on the right side (black arrow).

Color image of this figure appears in the color plate section at the end of the book.

can not be recommended. However, in some cases—dedicated stent types with a diameter of at least 3 mm—cardiac CT might be an alternative to invasive angiography.

MSCT Plaque Imaging

Detection and quantification of coronary calcifications as a risk marker for coronary events is a domain of cardiac CT (Schroeder et al. 2008). According to the 'St. Francis Heart Study', coronary calcifications are more accurate in the prediction of future events than standard risk factors or C-reactive protein (Arad et al. 2005). Coronary calcium is recommended to evaluate an individual's cardiovascular risk more accurately (De Backer et al. 2003). However, the absence of coronary calcifications does not necessarily mean the exclusion of (obstructive) CAD (Drosch et al. 2010).

Contrast-enhanced MSCT allows for the detection and classification of different stages of atherosclerosis, and coronary plaque burden can be assessed non-invasively with slight underestimation of non-calcified plaque volume in comparison to intravascular ultrasound (Brodoefel et al. 2008b). New semi-automatic post-processing software tools– based on attenuation of tissues—allow for more accurate and reproducible quantification of coronary plaque than visual assessment. However, plaque composition is still impossible to quantify by cardiac CT in comparison to virtual histology (Brodoefel et al. 2008b) (Fig. 6.3).

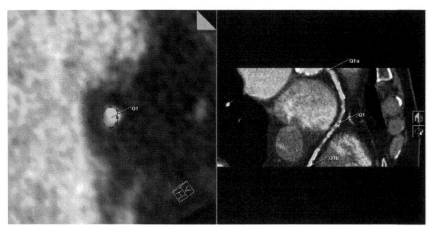

Figure 6.3 Plaque imaging with multi-slice computed tomography. Unpublished images.

Semiautomatic contour detection with post-processing software allows quantifying coronary plaque burden within the total vessel. On the left side cross-section of a coronary lesion ('Q1') which is located in the middle of the right coronary artery (right side). Different colours indicated different contrast attenuation of the plaque components.

Color image of this figure appears in the color plate section at the end of the book.

The impact of non-calcified plaques detected by MSCT is currently still unclear. It might be an additional risk factor for acute cardiac events or a target for treatment with lipid lowering drugs since a significant regression of non-calcified plaque burden (–24 % within one year) together with an increase of calcifications of + 17 % could be demonstrated (Burgstahler et al. 2007).

CT imaging of the Myocardium and the Cardiac Chambers

Contrast-enhanced MSCT can provide valuable information about myocardial perfusion. Myocardial perfusion defects might be observed in the early phase of the contrast bolus ('early defect'), with residual defects and late enhancement comparable to cardiac magnetic resonance tomography (CMR) (Koyama et al. 2004).

A good agreement between cardiac MSCT and CMR was demonstrated by Mahnken et al. (Mahnken et al. 2005) in 28 patients with reperfused myocardial infarction. Density values measured within infracted myocardium were significantly different from viable myocardium in early as well as in late phase CT. Contrast enhanced MSCT permitted assessment of acute myocardial infarction, and late enhancement in MSCT was as reliable as delayed contrast enhanced CMR in assessing the size of acute myocardial infarction and myocardial viability. However, up to now, cardiac CMR remains the non-invasive gold standard to assess the size of myocardial scars, myocardial viability and myocardial structure.

Cardiac CT to Detect Coronary Anomalies

Cardiac CT is helpful in diagnosing coronary anomalies since it allows the visualization of the coronary arteries in relation to other cardiac or vascular structures. This is of major importance as coronary anomalies may mimic coronary artery disease. Disorders like a left coronary artery originating from the right sinus of Valsalva running between aorta and pulmonary trunk or a Bland-White-Garland-Syndrome can easily be visualized (image example of a coronary anomaly, see Fig. 6.4).

Coronary anomalies might mimic coronary artery disease and can easily be evaluated by cardiac CT.

Cardiac CT and the Assessment of Left Ventricular Function

On commercially available workstations, functional parameters such as end-diastolic volume, end-systolic volume, stroke volume, ejection fraction and myocardial mass can be calculated from cardiac CT angiography datasets. Various studies have shown that for these left ventricular functional parameters MSCT of the heart with retrospective ECG-gating shows a high correlation with CMR (Schroeder et al. 2008) However, due to the radiation exposure

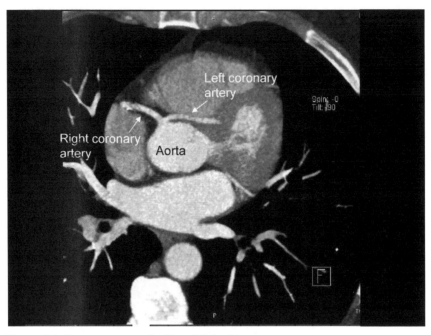

Figure 6.4 Coronary anomaly. Unpublished image.
This figure demonstrates a common ostium of the right and left coronary artery.
Color image of this figure appears in the color plate section at the end of the book.

and the need for contrast media cardiac CT for evaluation of left ventricular function should only be used in dedicated patients (e.g., poor image quality in echocardiography and contraindications for cardiac MRI).

Cardiac CT and the Assessment of Valve Disease

The assessment of aortic valve stenosis using multi-slice computed tomography is feasible with good diagnostic accuracy in comparison to transthoracic echocardiography (Feuchtner et al. 2007).

In a second study published by Feuchtner et al. the accuracy of cardiac CT to detect and quantify aortic valve insufficiency in comparison to TTE was tested (Feuchtner et al. 2006). A visible central valvular leakage area was considered to be a diagnostic criterion for aortic regurgitation. The overall sensitivity of MSCT for the identification of patients with aortic regurgitation was 81% with a positive predictive value of 95%. However, severe calcifications—which are more common in degenerative valvular disease—limited the diagnostic accuracy.

Willmann et al. published data from patients with mitral valve disease in which MSCT was performed. They found MSCT to be helpful in detecting valve abnormalities like thickening of the mitral valve leaflets, presence of mitral annulus calcification and calcification of the valvular leaflets (Willmann et al. 2002). However, up to now, there are no larger clinical trials comparing MSCT with echocardiography in terms of mitral valve regurgitation.

Cardiac CT allows visualizing cardiac valve morphology with high accuracy.

Extracardiac Findings

When performing a cardiac MSCT scan, the anatomic status of other thoracic organs should also be evaluated. According to a study published by Gil et al. a relevant number of patients undergoing cardiac CT have noncardiac disease (Gil et al. 2007). Out of 258 asymptomatic patients, 145 (56.2%) were found to have a significant noncardiac abnormality such as lung abnormalities (emphysema, bullae, interstitial lung disease, mass or nodules), pericardial abnormalities, liver disease, adrenal masses and bone abnormalities requiring additional workup. Hence, knowledge of extracardiac structures and pathology are mandatory when reading cardiac MSCT scans especially as some of these entities might be a differential diagnosis to coronary artery disease.

Cardiac CT-scans may discover non-cardiac disease. Hence, all scan should be evaluated for non-cardiac pathologies.

Radiation Exposure and dose Reduction

The application of X-rays must be considered the major limitation of cardiac CT (beside the need of iodinated contrast agent with possible side effects like renal insufficiency or induction of hyperthyroidism) and limits its use for routine screening purposes. A publication from Einstein et al. estimating the risk of cancer associated with 64-slice

MSCT report, e.g., the lifetime cancer risk for standard cardiac scans to be 1 in 143 for a 20-yr-old woman (Einstein et al. 2007). Due to the radiation exposure, MSCT of children, adolescent patients and younger persons should only be performed in particular cases, and other non-invasive imaging techniques such as ultrasound, echocardiography or magnetic resonance imaging should be preferred.

Dose reduction in cardiac CT

Standard techniques to reduce radiation exposure are dose modulation, lowering the tube voltage to 100 kV in non-obese patients and minimizing the scan length. Beyond these techniques, sequential scan mode with prospective electrocardiogram triggering might be considered in patients with a stable heart rate. This so called 'step-and-shoot'-technique has come into focus within the last 2 yr. Meanwhile non-invasive coronary angiography can be performed with an estimated effective dose of approximately 1 mSv (Achenbach et al. 2010), which is below the annual natural radiation exposure.

Using this technique, functional imaging (evaluation of left or right ventricular function, imaging of valve abnormalities) is, however, not possible as the heart is merely imaged at a dedicated time point during diastole.

The application of ionizing radiation is one important limitation of cardiac CT. However, up-to-date scanners allow visualizing of the coronary arteries with approximately 1 mSv.

Cost Effectiveness of Cardiac CT

Interestingly, cardiac CT proved to be the most cost-effective tool (in comparison to CMR and invasive angiography) in patients with a likelihood for coronary artery disease of up to 50%. At a pretest likelihood of 60% invasive angiography and cardiac CT were equally effective, while the invasive procedure was most effective in patients with a pretest likelihood of at least 70% (Dewey and Hamm 2007). Comparable data were recently published in terms of asymptomatic persons with positive stress tests (Halpern et al. 2010).

Despite these data, up to now cardiac CT is not reimbursed by most public insurances in most countries.

Cardiac CT is cost-effective in patients with a low- to intermediate risk of having coronary artery disease.

INDICATIONS FOR CARDIAC CT

The latest appropriateness criteria based on a review of literature for the American College of Cardiology have been published in 2010 (Taylor et al. 2010). Another paper concerning this issue was published in 2007 by the Working Group Nuclear Cardiology and Cardiac CT of the European Society of Cardiology and the European Council of Nuclear Cardiology (Schroeder et al. 2008). Indications currently judged to be appropriate to perform cardiac CT angiography are summarized in Table 6.2.

OUTLOOK

As outlined above, MSCT provides detailed morphological information about the coronary arteries and coronary plaques. However, 'ischemia' itself can not be visualized by MSCT. On the other hand, positron emission tomography is able to detect coronary ischemia but not to visualize coronary arteries. Combining these two image modalities ('hybrid imaging') allows a comprehensive imaging of cardiac function with anatomical coregistration. Javadi et al. could just recently demonstrate that functional and morphological heart imaging combining MSCT and PET is possible with less than 6 mSv (Javadi et al. 2008). With the growing numbers of hybrid scanners, PET-CT will no longer be restricted to patients with malignant diseases but will help to handle patients with cardiovascular disorders.

Another important issue reflects molecular imaging with MSCT which was initially restricted to scintigraphic methods. In 2007, Hyafil et al. were able to demonstrate that macrophages in atherosclerotic plaques can be detected with a clinical CT scanner after the intravenous injection of a contrast agent in rabbits (Hyafil et al. 2007). Although results in humans are still missing, these data

Table 6.2 Appropriate indications for cardiac CT coronary angiography, adapted from (Taylor et al. 2010). Unpublished table.

Detection of CAD: Symptomatic—Evaluation of Non-Acute Chest Pain Syndrome (Use of CT Angiogram)
• Intermediate pre-test probability of CAD
• ECG uninterpretable OR unable to exercise • low pre-test probability of CAD
• ECG uninterpretable OR unable to exercise • intermediate pre-test probability of CAD
Detection of CAD: Symptomatic—Evaluation of Intra-Cardiac Structures (Use of CT Angiogram)
• Evaluation of suspected coronary anomalies
Detection of CAD: Symptomatic—Acute Chest Pain (Use of CT Angiogram)
• Low or intermediate pre-test probability of CAD • ECG and cardiac biomarkers normal or equivocal
New-Onset or Newly Diagnosed Clinical Heart Failure and No Prior CAD
• Low or intermediate pre-test probability of CAD
Preoperative Coronary Assessment Prior to Noncoronary Cardiac Surgery
• Low or intermediate pre-test probability of CAD
Detection of CAD With Prior Test Results-Evaluation of Chest Pain Syndrome (Use of CT Angiogram)
• Uninterpretable or equivocal stress test (e.g. exercise, perfusion, stress echo)
Evaluation of New or Worsening Symptoms in the Setting of Past Stress Imaging Study
• Previous stress imaging study normal
Risk Assessment Postrevascularization (PCI or CABG)—Symptomatic
• Evaluation of graft patency after CABG • Prior left main coronary stent with stent diameter ≥3 mm
Structure and Function—Morphology (Use of CT Angiogram)
• Assessment of complex congenital heart disease including anomalies of coronary circulation, great vessels, and cardiac chambers and valves

This table summarizes current appropriate indications for cardiac CT.

CAD: coronary artery disease: PCI: percutaneous coronary intervention; CABG: coronary artery bypass graft.

warrant the assumption that molecular imaging with MSCT will play an important role in the near future.

KEY FACTS OF CARDIAC COMPUTED TOMOGRAPHY

- Imaging of the coronary arteries non-invasively by the use of multi-slice computed tomography was first described at the end of the last century.
- As coronary arteries are of small dimensions and as they move rapidly, a high spatial as well as temporal resolution is mandatory for coronary imaging.
- The gantry rotation time of up-to-date scanners is approximately 330 ms. That means, the radiation source and detector rotates more than three times per second.
- Reading and interpreting non-invasive coronary angiography requires fundamental training in cardiac CT.
- Cardiac computed tomography is one of the most important technical advances in the care of cardiovascular patients within the last years.

SUMMARY

- Cardiac CT is still a rapidly evolving technology in non-invasive imaging of heart disease.
- In the meantime, there is evidence, that CT might be useful for risk assessment (coronary calcium), as well as for the non-invasive evaluation of the presence of coronary lesions in selected patients with CAD and a low or intermediate pre-test probability.
- Stent imaging with cardiac CT is only appropriate to evaluate left main stents in asymptomatic patients.
- Beside the visualization of coronary arteries, information about cardiac function, heart valves and cardiac morphology can be obtained by cardiac CT.
- Each scan should be evaluated for non-cardiac disease as many patients have accompanying (pathological) findings.
- Cardiac CT is cost-effective in patients with a low or intermediate pre-test probability of having coronary artery disease.
- Technical improvement and especially the efforts to reduce radiation exposure will expand the indication for cardiac CT in clinical cardiology within the next years.

DEFINITIONS

Coronary anomaly: At least one of the coronary arteries has an atypical course. This might lead to chest pain even in younger persons.

Pre-test probability: That means, the probability of having a coronary artery stenosis in a patient prior to cardiac CT.

Cost-effectiveness: Is it worth spending money for a cardiac CT as some more expensive procedures might be cancelled depending on the CT result?

Post-processing: Raw data are transferred to an 'off-line' workstation for further evaluation of the CT scan (e.g., 3-D-reconstruction, evaluation of left ventricular function).

Hybrid-imaging: Two different image modalities (e.g., CT and PET or MRI and PET) are performed and the respective images are merged to only one illustration. By the use of this technique functional information of MRI or PET and anatomical information from CT can be correlated ('which stenosis is responsible for ischemia').

LIST OF ABBREVIATIONS

CAD	:	Coronary Artery Disease
CMR	:	Cardiac Magnetic Resonance Imaging
CT	:	Computed Tomography
DSCT	:	Dual Source Computed Tomography
MSCT	:	Multi Slice Spiral Computed Tomography

REFERENCES CITED

Achenbach, S., D. Ropers, A. Kuettner, T. Flohr, B. Ohnesorge, H. Bruder, H. Theessen, M. Karakaya, W.G. Daniel, W. Bautz, W.A. Kalender and K. Anders. 2006. Contrast-enhanced coronary artery visualization by dual-source computed tomography—initial experience: Eur J Radiol vol. 57, pp. 331–335.

Achenbach, S., M. Marwan, D. Ropers, T. Schepis, T. Pflederer, K. Anders, A. Kuettner, W.G. Daniel, M. Uder and M.M. Lell. 2010. Coronary computed tomography angiography with a consistent dose below 1 mSv using prospectively electrocardiogram-triggered high-pitch spiral acquisition: Eur. Heart J vol. 31, pp. 340–346.

Arad,Y., K.J. Goodman, M. Roth, D. Newstein, A.D. Guerci. 2005. Coronary calcification, coronary disease risk factors, C-reactive protein, and atherosclerotic cardiovascular disease events: the St. Francis Heart Study: J Am Coll Cardiol vol. 46, pp. 158–165.

Brodoefel, H., C. Burgstahler, I. Tsiflikas, A. Reimann, S. Schroeder, C.D. Claussen, M. Heuschmid and A.F. Kopp. 2008a. Dual-source CT: effect of heart rate, heart rate variability, and calcification on image quality and diagnostic accuracy: Radiology vol. 247, pp. 346–355.

Brodoefel,H., A. Reimann, M. Heuschmid, I. Tsiflikas, A.F. Kopp, S. Schroeder, C.D. Claussen, M.E. Clouse and C. Burgstahler. 2008b. Characterization of coronary atherosclerosis by dual-source computed tomography and HU-based color mapping: a pilot study: Eur Radiol vol. 18, pp. 2466–2474.

Burgstahler, C., A. Reimann, T. Beck, A. Kuettner, D. Baumann, M. Heuschmid, H. Brodoefel, C.D. Claussen, A.F. Kopp and S. Schroeder. 2007. Influence of a lipid-lowering therapy on calcified and noncalcified coronary plaques monitored by multislice detector computed tomography: results of the New Age II Pilot Study: Invest Radiol vol. 42, pp. 189–195.

De Backer, G., E. Ambrosioni, K. Borch-Johnsen, C. Brotons, R. Cifkova, J. Dallongeville, S. Ebrahim, O. Faergeman, I. Graham, G. Mancia, C. Manger, V, K. Orth-Gomer, J. Perk, K. Pyorala, J.L. Rodicio, S. Sans, V. Sansoy, U. Sechtem, S. Silber, T. Thomsen and D. Wood. 2003. European guidelines on cardiovascular disease prevention in clinical practice. Third Joint Task Force of European and Other Societies on Cardiovascular Disease Prevention in Clinical Practice: Eur Heart J vol. 24, pp. 1601–1610.

Dewey, M. and B. Hamm. 2007. Cost effectiveness of coronary angiography and calcium scoring using CT and stress MRI for diagnosis of coronary artery disease: Eur Radiol vol. 17, pp. 1301–1309.

Drosch, T., H. Brodoefel, A. Reimann, C. Thomas, I. Tsiflikas, M. Heuschmid, S. Schroeder and C. Burgstahler. 2010. Prevalence and clinical characteristics of symptomatic patients with obstructive coronary artery disease in the absence of coronary calcifications: Acad Radiol vol. 17, pp. 1254–1258.

Einstein, A.J., M.J. Henzlova and S. Rajagopalan. 2007. Estimating risk of cancer associated with radiation exposure from 64-slice computed tomography coronary angiography: JAMA vol. 298, pp. 317–323.

Feuchtner, G.M., W. Dichtl, T. Schachner, S. Muller, A. Mallouhi, G.J. Friedrich and D.Z. Nedden. 2006. Diagnostic performance of MDCT for detecting aortic valve regurgitation: AJR Am J Roentgenol vol. 186, pp. 1676–1681.

Feuchtner, G.M., S. Muller, J. Bonatti, T. Schachner, C. Velik-Salchner, O. Pachinger, W. Dichtl. 2007. Sixty-four slice CT evaluation of aortic stenosis using planimetry of the aortic valve area: AJR Am J Roentgenol vol. 189, pp. 197–203.

Gil, B.N., K. Ran, G. Tamar, F. Shmuell and A. Eli. 2007. Prevalence of significant noncardiac findings on coronary multidetector computed tomography angiography in asymptomatic patients: J Comput Assist Tomogr vol. 31, pp. 1–4.

Halpern, E.J., M.P. Savage, D.L. Fischman and D.C. Levin. 2010. Cost-effectiveness of coronary CT angiography in evaluation of patients without symptoms who have positive stress test results: AJR Am J Roentgenol vol. 194, pp. 1257–1262.

Hyafil, F., J.C. Cornily, J.E. Feig, R. Gordon, E. Vucic, V. Amirbekian, E.A. Fisher, V. Fuster, L.J. Feldman and Z.A. Fayad. 2007. Noninvasive detection of macrophages using a nanoparticulate contrast agent for computed tomography: Nat Med vol. 13, pp. 636–641.

Javadi, M., M. Mahesh, G. McBride, C. Voicu, W. Epley, J. Merrill and F.M. Bengel. 2008. Lowering radiation dose for integrated assessment of coronary morphology and physiology: first experience with step-and-shoot CT angiography in a rubidium 82 PET-CT protocol: J Nucl Cardiol vol. 15, pp. 783–790.

Koyama, Y., T. Mochizuki and J. Higaki. 2004. Computed tomography assessment of myocardial perfusion, viability, and function: J Magn Reson Imaging vol. 19, pp. 800–815.

Mahnken, A.H., R. Koos, M. Katoh, J.E. Wildberger, E. Spuentrup, A. Buecker, R.W. Gunther and H.P. Kuhl. 2005. Assessment of myocardial viability in reperfused acute myocardial infarction using 16-slice computed tomography in comparison to magnetic resonance imaging: J Am Coll Cardiol vol. 45, pp. 2042–2047.

Mowatt, G., E. Cummins, N. Waugh, S. Walker, J. Cook, X. Jia, G.S. Hillis and C. Fraser. 2008. Systematic review of the clinical effectiveness and cost-effectiveness of 64-slice or higher computed tomography angiography as an alternative to invasive coronary angiography in the investigation of coronary artery disease: Health Technol Assess vol. 12, pp. iii–143.

Nasis, A., M.C. Leung, P.R. Antonis, J.D. Cameron, S.J. Lehman, S.A. Hope, M.P. Crossett, J.M. Troupis, I.T. Meredith and S.K. Seneviratne. 2010. Diagnostic accuracy of noninvasive coronary angiography with 320-detector row computed tomography: Am J Cardiol vol. 106, pp. 1429–1435.

Nikolaou, K., A. Knez, C. Rist, B.J. Wintersperger, A. Leber, T. Johnson, M.F. Reiser and C.R. Becker. 2006. Accuracy of 64-MDCT in the diagnosis of ischemic heart disease: AJR Am J Roentgenol vol. 187, pp. 111–117.

Oncel, D., G. Oncel and A. Tastan. 2007. Effectiveness of dual-source CT coronary angiography for the evaluation of coronary artery disease in patients with atrial fibrillation: initial experience: Radiology vol. 245, pp. 703–711.

Ong, T.K., S.P. Chin, C.K. Liew, W.L. Chan, M.T. Seyfarth, H.B. Liew, A. Rapaee, Y.Y. Fong, C.K. Ang and K.H. Sim. 2006. Accuracy of 64-row multidetector computed tomography in detecting coronary artery disease in 134 symptomatic patients: influence of calcification: Am Heart J vol. 151, pp. 1323–1326.

Ropers, D., J. Rixe, K. Anders, A. Kuttner, U. Baum, W. Bautz, W.G. Daniel and S. Achenbach. 2006. Usefulness of multidetector row spiral computed tomography with 64- x 0.6-mm collimation and 330-ms rotation for the noninvasive detection of significant coronary artery stenoses: Am J Cardiol vol. 97, pp. 343–348.

Ropers, U., D. Ropers, T. Pflederer, K. Anders, A. Kuettner, N.I. Stilianakis, S. Komatsu, W. Kalender, W. Bautz, W.G. Daniel and S. Achenbach. 2007. Influence of heart rate on the diagnostic accuracy of dual-source computed tomography coronary angiography: J Am Coll Cardiol vol. 50, pp. 2393–2398.

Schroeder, S., S. Achenbach, F. Bengel, C. Burgstahler, F. Cademartiri, P. de Feyter, R. George, P. Kaufmann, A.F. Kopp, J. Knuuti, D. Ropers, J. Schuijf, L.F. Tops and J.J. Bax. 2008. Cardiac computed tomography: indications, applications, limitations, and training requirements: report of a Writing Group deployed by the Working Group Nuclear Cardiology and Cardiac CT of the European Society of Cardiology and the European Council of Nuclear Cardiology: Eur Heart J vol. 29, pp. 531–556.

Taylor, A.J., M. Cerqueira, J.M. Hodgson, D. Mark, J. Min, P. O'Gara, G.D. Rubin, C.M. Kramer, D. Berman, A. Brown, F.A. Chaudhry, R.C. Cury, M.Y. Desai, A.J. Einstein, A.S. Gomes, R. Harrington, U. Hoffmann, R. Khare, J. Lesser, C. McGann, A. Rosenberg, R. Schwartz, M. Shelton, G.W. Smetana and S.C. Smith, Jr. 2010. ACCF/SCCT/ACR/AHA/ASE/ASNC/NASCI/SCAI/SCMR 2010 appropriate use criteria for cardiac computed tomography: a report of the American College of Cardiology Foundation Appropriate Use Criteria Task Force, the Society of Cardiovascular Computed Tomography, the American College of Radiology, the American Heart Association, the American Society of Echocardiography, the American Society of Nuclear Cardiology, the North American Society for Cardiovascular Imaging, the Society for Cardiovascular Angiography and Interventions, and the Society for Cardiovascular Magnetic Resonance: J Am Coll Cardiol vol. 56, pp. 1864–1894.

Tsiflikas, I., T. Drosch, H. Brodoefel, C. Thomas, A. Reimann, A. Till, D. Nittka, A.F. Kopp, S. Schroeder, M. Heuschmid and C. Burgstahler. 2009. Diagnostic accuracy and image quality of cardiac dual-source computed tomography in patients with arrhythmia: Int J Cardiol.

Voros, S. 2009. What are the potential advantages and disadvantages of volumetric CT scanning?: J Cardiovasc Comput Tomogr vol. 3, pp. 67–70.

Willmann, J.K., R. Kobza, J.E. Roos, M. Lachat, R. Jenni, P.R. Hilfiker, T.F. Luscher, B. Marincek and D. Weishaupt. 2002. ECG-gated multi-detector row CT for assessment of mitral valve disease: initial experience: Eur Radiol vol. 12, pp. 2662–2669.

7

Drugs Used in Angina: An Overview

Mario Marzilli[1,*] and *Alda Huqi*[1,2]

ABSTRACT

As compared to medical therapy, revascularization has not been shown to significantly affect morbidity and mortality. In fact, following revascularization a considerable number of patients continue to experience angina and therefore also require medical therapy. Revascularization carries, although minimal, intra-procedural risks for adverse events and is not equally feasible in all parts of the world. For this reason, drug therapy is still regarded as an undisputed masterpiece in the management of angina patients. Nitrates, beta-blockers (BBs) and calcium channel blockers (CCBs) are the traditional (hemodynamic) anti-anginal drugs. However, results are not yet optimal. Neither revascularization, nor classical anti-anginal drugs (and a various combinations of them) confer angina relief in all patients or a reduction in cardiovascular events. Such disappointing results have led to the development of non-hemodynamic agents that show additional benefits. A tailored symptomatic therapy together with an overall risk reduction therapy is currently the best treatment option for reducing morbidity and mortality from IHD.

---

[1]Cardio-Thoracic and Vascular Department, Via Paradisa, 2, 56100–Pisa, Italy; Email: mario.marzilli@med.unipi.it

[2]Mazankowski Alberta Heart Institute, T6G 2S2, Edmonton, Alberta, Canada; Email: alda_h@hotmail.com; huqi@ualberta.ca

*Corresponding author

List of abbreviations after the text.

INTRODUCTION

Ischemic heart disease (IHD) is a leading cause of mortality and stable angina represents the most frequent clinical presentation (Gibbons et al. 2003). Besides being associated with high morbidity and mortality, angina symptoms are also well known to be associated with depression and poor quality of life (Rumsfeld et al. 2003). Antianginal treatment has not been shown to modify the natural history of disease. Therefore, the primary goal in the management of such patients is symptom control and amelioration of quality of life. Revascularization by means of coronary artery bypass graft (CABG) surgery or percutaneous coronary intervention (PCI), drug therapy and a various combination of such options represent the available treatment strategies for patients with stable coronary artery disease (CAD).

In this chapter we will focus on traditional medical therapy and the rationale for use of novel anti-anginal drugs. However, since revascularization often accompanies, follows or even precedes implementation of medical therapy, relative superiority and indications will also be discussed. In addition, we will also provide an overview of global management of patients with stable angina pectoris.

MEDICAL THERAPY VERSUS REPERFUSION STRATEGY

Chronic stable angina is the result of a precarious balance between oxygen supply and requirements of the myocardium, which gets clinically manifested in conditions of disturbed equilibrium such as physical activity (i.e., effort angina). Accordingly, interventions aimed at either increasing oxygen supply or decreasing oxygen requirement increase the threshold for myocardial ischemia, and therefore result in amelioration of symptoms.

Myocardial revascularization (either percutaneous or surgical) in chronic angina patients aims at the local removal or bypass of the epicardial obstacle to the coronary flow. This approach is in agreement with the main proposed pathophysiological mechanism for angina: reduced oxygen supply due to epicardial coronary stenosis. However, stenosis removal does not always result in complete and permanent relief of symptoms. In fact, after coronary revascularization, a considerable number of patients continue to complain of symptoms and as many as two thirds might require one or more anti-angina agents (Holubkov et al. 2002). In addition, despite being usually more effective in symptom control, treatment with CABG or PCI of patients with chronic CAD has not been shown to result in significant improvement in survival or decreased incidence of acute coronary events when compared to medical treatment (Hueb et al. 2004). These findings were also confirmed in the COURAGE trial (Boden et al. 2007) which is regarded as practice changing and a basis for recommending optimal medical therapy (OMT) as initial therapy for stable angina. The COURAGE trial also indicates that in stable patients, deferring intervention while under OMT is a viable approach and does not significantly raise risk. Moreover, revascularization carries, although minimal, intra-procedural risks for adverse events and may be too costly or inaccessible for many patients in developing countries. For this reason, drug therapy is still regarded as an undisputed masterpiece in the management of angina patients, with revascularization being restrained to cases of persistent symptoms despite OMT.

Lifestyle changes and OMT have been underestimated and are more powerful than previously believed. Nonetheless, despite such evidence, a significant proportion of patients with chronic stable angina undergoing elective PCI do not receive adequate recommended anti-anginal therapy (Elder et al. 2010) prior to revascularization.

CLASSICAL ANTI-ANGINA DRUGS

Classical anti-anginal drug therapy is based on treatment with nitrates, beta blockers (BBs), and calcium channel blockers (CCBs). These are the so called 'hemodynamic agents' which act by lowering rate-pressure product and/or producing systemic venodilation, thereby lowering left ventricular end-diastolic pressure and volume and reducing myocardial wall tension. Although their main action is a reduction in oxygen requirements, a decrease in myocardial wall tension also permits a greater flow in the epicardial coronary arteries and improves myocardial oxygen delivery.

NITRATES

The predominant effect of nitrates is to reduce preload, exerting a greater activity in the venous than arterial beds. However, at higher doses a direct effect upon arteries also becomes evident, resulting in a reduction in blood pressure (BP) and afterload. These effects translate into reduced myocardial oxygen consumption and a higher threshold level before angina is triggered.

All patients with angina symptoms should be given sublingual nitroglycerine. Efficacy duration is on the order of 30 min. Patients with stable chronic angina should use nitroglycerin prophylactically about 5 min prior to any stress or activity that is known to trigger symptoms, as well as for acute events. Side effects include cerebral vasodilation with headache, postural hypotension, dizziness, and rarely, syncope in hypovolemic patients.

Long-acting nitrates are often prescribed as prophylactic antianginal drugs for chronic therapy. However, except for isosorbide mononitrate, they are subjected to hepatic first-pass metabolism and, therefore, besides the need for higher doses, their efficacy is also less predictable. These agents effectively extend the duration of action of sublingual nitroglycerin, but none provide full 24-hr protection (Thadani 1997). In fact, the biggest issue with the use of long-acting nitrates is development of tolerance within 12–24 hr, which may however be avoided with a nitrate free period of about 8 hr each day or up to three times a day per oral administration.

A recent meta-analysis of randomized controlled trials showed that both continuous and intermittent nitrate therapy were efficacious in reducing angina attacks and prolonging exercise duration. Nonetheless, these effects did not translate in an amelioration of quality of life and, moreover, about half of patients complained of a headache (Wei et al. 2010).

BETA-BLOCKERS

BBs inhibit the action of endogenous catecholamines (epinephrine and norepinephrine in particular) on adrenergic receptors, thereby inhibiting calcium influx into the cell. Their anti-anginal effects are mediated through a reduction in ventricular inotropy, heart rate and a decrease in the maximal velocity of myocardial fiber shortening, therefore keeping myocardial oxygen demand below the threshold at which angina occurs. Different BBs differ in their effects on the three adrenergic receptors (β1, β2, and α) and in effect duration (Table 7.1). Common adverse drug reactions are: nausea, diarrhea, bronchospasm, dyspnea, cold extremities, exacerbation of Raynaud's syndrome, bradycardia, hypotension, heart failure (HF), heart block, fatigue, dizziness, abnormal vision, decreased concentration, hallucinations, insomnia, nightmares, depression, erectile dysfunction and alteration of glucose and lipid metabolism.

Table 7.1 Pharmacological characteristics of principal BBs used in clinical practice.

Drug	Daily dosing (mg/d)	Frequency	Half-life (hr)	Cardio-selective	Partial agonist activity	Alfa antagonist activity
Atenolol	50–100	Once	6–9	yes	no	No
Bisoprolol	5–20	Once	9–12	yes	no	No
Carvedilol	12.5–50	Twice	7–10	no	no	Yes
Carvedilol phosphate	20–80	Once	10.6–11.5	no	no	Yes
Labetalol	200–1200	Twice	3–6	no	no	Yes
Metoprolol tartrate	50–200	Twice	3–7	yes	no	No
Metoprolol succinate	50–400	Once	3–7	yes	no	No
Nadolol	20–240	Once	10–20	no	no	No
Nebivolol	5–40	Once	12–19	yes	no	No
Pindolol	10–60	Twice	3–4	no	yes	No
Propanolol	40–240	Twice	3–4	no	no	No
Propanolol long-acting	60–240	Once	8–11	no	no	No

Propranolol is the first BB shown to improve symptoms in patients with angina pectoris in a multicenter controlled trial (Grant et al. 1966). However, at comparable doses, the majority of head-to-head BBs trials have shown no major differences in terms of exercise tolerance, attack frequency, or nitroglycerin use (Kardas 2007). Importantly, while their beneficial effect on outcomes in patients who have experienced a myocardial infarction (MI) has been validated in a number of clinical studies, their impact on 'pure' stable angina patients has been shown to be less clinically incisive. As a matter of fact, studies that have evaluated the efficacy of BBs treatment in stable angina have included only 'soft outcomes' such as exercise tolerance, attack frequency and nitrate use. According to current guidelines, unless contraindications exist, beta blockers have a class I (level of evidence: A) indication for first-line therapy in patients with angina who have suffered a previous MI (Fraker et al. 2007).

CALCIUM CHANNEL BLOCKERS

CCBs are potent coronary and systemic arterial vasodilators that reduce BP as well as cardiac contractility. They lower the frequency of angina, reduce the need for nitrates, extend treadmill walking time, and improve ischemic ST-segment changes on exercise testing and electrocardiographic monitoring.

There are three classes of CCBs (Table 7.2). Dihydropyridines show a high vascular L-type calcium channel selectivity and therefore are primarily used to treat hypertension. They are not commonly used to treat angina because their powerful systemic vasodilator and BP lowering effects can lead to reflex cardiac stimulation (tachycardia and increased inotropy), which can dramatically increase myocardial oxygen demand. Non-dihydropyridines include verapamil (phenylalkylamine class) and diltiazem (benzothiazepine class) which have more selective cardio-inhibitory properties (Fig. 7.1). Verapamil is relatively selective for the myocardium, and is less effective as a systemic vasodilator drug. This drug has a very important role in treating angina (by reducing myocardial oxygen demand and reversing coronary vasospasm) and arrhythmias. Diltiazem is an intermediate between verapamil and

Table 7.2 Differences of tissue selectivity between dihydropiridines (nifedipine and others), diltiazem and verapamil.

Drug category	Peripheral and coronary vasodilation	Depression of cardiac contractility	Depression of SA node	Depression of AV node
Nifedipine (dihydropiridine)	+++++	+	+	0
Diltiazem (benzothiazepine)	+++	++	+++++	++++
Verapamil (phenylalkylamine)	++++	++++	+++++	+++++

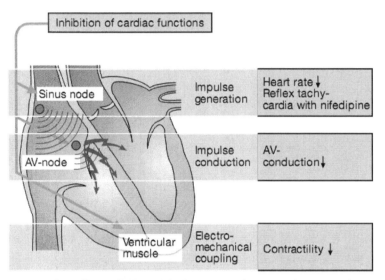

Figure 7.1 Cardiac effects of CCBs. CCBS cause a reduction in heart rate (except for dihydropiridines which may cause reflex tachycardia), atrio-ventricular conduction velocity and cardiac inotropy. Abbreviations: CCBs (calcium channel blockers), AV (atrio-ventricular).

Color image of this figure appears in the color plate section at the end of the book.

dihydropyridines in its selectivity for vascular calcium channels. By having both cardiac depressant and vasodilator actions, diltiazem is able to reduce arterial pressure without producing the same degree of reflex cardiac stimulation caused by dihydropyridines.

Common side effects such as headache, dizziness, flushing, and edema are due to vasodilation. Interactions with other negative chronotropic or inotropic agents to produce bradycardia, heart block, or HF have been reported. Therefore, patients having preexistent bradycardia, conduction defects, or HF caused by systolic dysfunction should not be given CCBs, especially the non-dihydropyridines.

Clinical efficacy of CCBs in angina patients is comparable to that of BBs. However, although CCBs are known to be effective antianginal agents, in agreement with other anti-anginal drugs, they do not modify the natural progression of the disease.

In conclusion, although effective, none of these drugs has been shown to be disease modifying—their use does not change the risk of MI, sudden cardiac death, or all-cause mortality. Moreover a combination therapy is often necessary for adequate symptom control. However, even when this strategy is adopted, a complete symptom relief may often not be the case. In one analysis, use of all three traditional agents still resulted in an average residual of two attacks of angina per week among participants (Pepine et al. 1994). Such overall unsatisfactory results have been the main drive for the development of alternative anti-anginal drugs. In fact, in many countries, ivabradine, nicorandil, ranolazine and trimetazidine constitute additional treatment options for angina patients.

ALTERNATIVE DRUG THERAPIES

Trimetazidine

Trimetazidine is a metabolic agent that inhibits fatty acid oxidation, thereby increasing glucose oxidation through an interdependence mechanism, called the 'Randle cycle'. In this way the ischemic myocardium can more efficiently use the limited oxygen supply which clinically translates into an amelioration of anginal symptoms. Trimetazidine has been used in Europe for at least two decades and its efficacy in alleviating chronic angina has been shown in several randomized clinical trials. Details on mechanism of action, clinical indications, efficacy and adverse effects are provided in an addressed chapter of this book.

Ranolazine

In 2006, ranolazine was approved in the U.S. for the relief of angina in patients who remained symptomatic on BBs, CCB, or nitrates (Chaitman 2006). It is a piperazine structurally related to trimetazidine, which was shown to have antischemic properties through promotion of glucose oxidation at the expense of fatty acid oxidation since the early 90s (Fig. 7.3) (McCormack et al. 1996). However, additional properties such as reduction in intracellular calcium overload through inhibition of the late sodium channels have recently gained more attention (Fig. 7.4) (Wasserstrom et al. 2009). These effects have been associated with a preserved mitochondrial structure, decreased intracellular calcium content and, finally, decreased post-ischemic ventricular fibrillation, myocardial stunning and infarct size. For this reason ranolazine is currently considered a first generation of a new drug category (i.e., inhibitor of late sodium currents). Nonetheless, it's important to note that therapeutic concentrations at which a reduction in calcium overload is observed are similar to those at which an increase in glucose oxidation has been documented (McCormack et al. 1996). However, regardless of the action mechanism, ranolazine has been shown to confer significant clinical benefit in angina patients. Its use is associated with a significant prolongation in exercise duration to angina and to ST-segment depression (1 mm), either in comparison or on top of defined anti-angina drugs (Rousseau et al. 2005). Nonetheless, its effects on the morbidity of angina patients remain still to be determined. Common side effects associated with its use include dizziness and constipation.

Ivabradine

In 2005, ivabradine was approved as an anti-anginal drug in Europe. This novel drug selectively inhibits the hyperpolarization activated (mixed sodium/potassium) inward I_f current, which is a primary sinoatrial node (SAN) pacemaker current, and therefore decreases rest and exercise heart rate responsiveness (Fig. 7.2). Importantly, Ivabradine does not affect contractility or atrio-ventricular nodal conduction, nor alter hemodynamics. The most frequent adverse

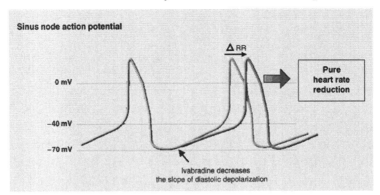

Figure 7.2 Effects of ivobradine of heart rate. Ivabradine slows spontaneous activity of the senoatrial node (SAN) by blocking I_f–channels, thereby causing a decreased rate of diastolic depolarization (ΔR). This effect is not associated by a decrease in cardiac inotropy and therefore constitutes a pure heart rate decrease.

Color image of this figure appears in the color plate section at the end of the book.

effect (<15%) is a transient brightness in the visual fields because the drug also blocks a retinal current with similar characteristics (I_h channels). Other adverse reactions, including conduction abnormalities, occur in less than 10% of cases.

In noninferiority trials, ivabradine compared well to atenolol or amlodipine (Ruzyllo et al. 2007). Importantly, its use in association with BBs has recently been shown to be both safe and efficacious in obtaining a better control of angina symptoms (Koester et al. 2010). However, benefits are usually confined to patients with higher resting cardiac frequency. At this time, ivabradine has not yet been approved in the U.S.

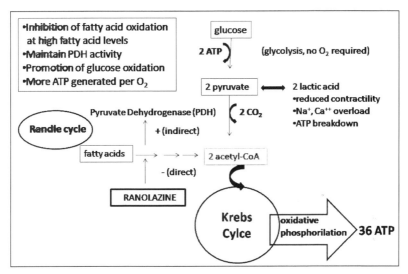

Figure 7.3 Ranolazine inhibits fatty acid oxidation, therefore causing a relief in the inhibition of PDH, the rate limiting enzyme of glucose oxidation. This effect causes an increase in glucose oxidation rates and therefore reduced proton accumulation, reduced calcium overload and increased contractility. Abbreviations: PDH (pyruvate dehydrogenase), ATP (adenosine triphosphate), CO_2 (carbon dioxide), Na^+ (sodium ions), Ca^{++} (calcium ions).

Nicorandil

Nicorandil is structurally a nicotinamide derivative with a nitrate moiety and a dual mechanism of action. First, it increases potassium ion conductance by opening adenosine triphosphate (ATP)-sensitive potassium channels, in turn activating the enzyme guanylate cyclase.

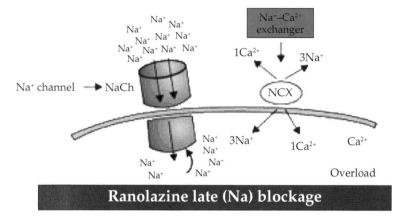

Figure 7.4 Ranolozine as a late sodium channel inhibitor. Inhibition of late sodium channels by ranolazine is compensated by the sodium-calcium exchanger channels. For each three molecules of sodium transported inside by this channel, a molecule of calcium is transported outside the cell, in this way, reducing calcium overload. Abbreviations: Na (sodium), Ca (calcium), NaCh (sodium channels), NCX (sodium calcium exchanger).

Second, nicorandil shares the smooth muscle-relaxing property of nitrates to vasodilate, lowering preload through venodilation. The drug also reduces afterload and promotes expression of endothelial NO synthase (Jahangir et al. 2001). In chronic angina patients, nicorandil has been shown to prolong time to the onset of angina and ischemic ECG changes, to extend exercise duration (Markham et al. 2000), and reverse ischemia-related impairment in regional wall motion. Moreover, in the Impact Of Nicorandil in Angina (IONA) study of 5,126 patients with angina (Effect of nicorandil on coronary events in patients with stable angina: the Impact Of Nicorandil in Angina (IONA) randomised trial 2002), nicorandil produced a significant reduction in hospitalization for chest pain, MI, and death. These effects have recently been ascribed to its anti-atherogenic effects, which seem to be different from that of statin therapy. However results have not always been consistent and, its use has been associated with induction of severe perianal ulcerations. This drug is not yet approved for use in the U.S., but it is available in other countries.

Rho-kinase Inhibitors

Rho kinase, or "ROCK", is an important intracellular enzyme which phosphorylates proteins to affect a number of cellular functions, among which myosin, resulting in smooth muscle contraction and vasoconstriction. In animal models of IHD, inhibition of myosin phosphorylation by fasudil resulted in a decrease of vascular smooth muscle hypercontraction and reduction of ischemic ST segment depression. These effects were reproduced in a phase II, double-blind, placebo controlled, randomized clinical trial of 84 patients with chronic angina (Vicari et al. 2005). Fasudil has also been shown to induce a further dilation at the site of coronary spasm in patients with vasospastic angina, who had already been treated with nitroglycerin (Otsuka et al. 2008). However, despite the proven clinical efficacy and safety in angina patients, the drug is not yet commercially available for use.

OVERALL CARDIOVASCULAR RISK REDUCTION

After quality of life improvement, the second major goal in chronic angina patients is an overall cardiovascular risk reduction. Aggressive and tailored risk factor control with appropriate diet, exercise, treatment of diabetes, hypertension, dyslipidemia and smoking cessation should be adopted in all patients.

Lifestyle therapy is an efficacious, widely available, innocuous, and inexpensive form of management of patients with IHD. However, given difficulties such as lack of gratification due to 'unsavory' new lifestyle and lack of immediately visible benefits, this approach is unfortunately hardly adoptable and therefore underused in clinical practice.

Besides lifestyle changes, medical therapy with antiplatelet agents, statins, and angiotensin-converting enzyme inhibitors (ACE-I) are other available tools for reducing risk for cardiovascular morbidity/mortality.

Given its proven benefits in reducing the incidence of cardiovascular events, aspirin should be started at 75 to 162 mg per day and continued indefinitely in all patients, unless contraindicated (Class I, Level of evidence: A) (Fraker et al. 2007).

Although ineffective in alleviating angina symptoms, ACE-Is have been shown to be cardioprotective in patients with CAD (Pepine et al. 2003). An ACE-I should be considered in patients with stable CAD and angina pectoris who are at lower-risk (i.e., mildly reduced or normal left ventricular ejection fraction in whom cardiovascular risk factors are well controlled and revascularization has been performed) (Class IIa, Level of evidence: B) (Fraker et al. 2007).

Statins are another drug category with proven benefits in chronic angina patients. In fact it is recommended that all patients with chronic stable angina be treated with a statin to a goal of LDL-C of <70 mg/dl (Class IIa, Level of evidence: A) (Fraker et al. 2007).

In conclusion, myocardial ischemia is universally accepted to be the result of an imbalance between oxygen supply and requirements to the myocardium. The presence of flow limiting coronary stenosis is the main recognized pathological mechanism underlying this condition. While revascularization procedures are performed with the aim to remove the flow limiting stenosis, traditional medical therapy with hemodynamic agents aims at reducing oxygen demand of the myocardium. However, although effective, none of these treatment strategies or their combination confers symptomatic relief in all patients. Such incomplete benefits have led to the development of new, non hemodynamic agents that have been shown to have additional anti-angina properties. A tailored symptomatic therapy together with overall risk reduction therapy is currently the best treatment option for reducing the morbidity and mortality from IHD.

PRACTICE AND PROCEDURES

Evaluation of Quality of Life in Angina Patients

Seattle Angina Questionnaire (SAQ) is one of the most frequently used and widely validated clinical tool for quality of life assessment. It is a 19-item self-administered questionnaire that evaluates five features of disability due to chronic ischemic heart disease over the preceding 4 wk: physical limitation (question 1); angina stability (question 2); angina frequency (questions 3–4); treatment satisfaction (questions 5–8); well-being (questions 9–11). It is scored by assigning each response an ordinal value, beginning with 1 for the response that implies the lowest level of functioning. Results of the same scale are then summed up and converted in a score system, ranging from 0 to 100, with higher scores indicating better health status. Moreover, each feature has predefined threshold levels for detecting clinically important differences (CIDs).

During clinical evaluation for diagnosis and/or follow-up, angina patients should routinely be administered health status evaluation questionnaires. Such practice provides a more objective assessment for evaluation of response to therapy and can therefore guide treatment optimization.

KEY FACTS

1. **Contraindications to nitrate use:**
   - hypotension/cardiogenic shock;
   - concomitant use of phosphodiesterase inhibitors (PDE-5) (sildenafil, tadalafil, verdanafil);
   - previous history of syncope with concomitant use of nitrates;
   - patients with elevated intracranial pressure;
   - patients with hypertrophic obstructive cardiomyopathy.

2. **What else to know about BBs use in angina patients:**
   - Most anti-angina effects of BBs result from β1 inhibition.
   - BBs with intrinsic sympathomimetic activity are not used for angina patients.
   - Most BBs are well absorbed.
   - Lipid-soluble BBs (i.e., propranolol and metoprolol) have shorter half-lives because they are metabolized by the liver.
   - Hydrophilic BBs (i.e., atenolol and nadolol) are eliminated by the kidney and have longer half-lives.
   - In larger doses, β1-blockers may lose some specificity and inhibit β2-receptors.
   - Importantly, since β-adrenergic receptors may be up-regulated when patients are treated with BBs, these agents should not be abruptly discontinued, lest rebound vasoconstriction precipitate unstable angina or even MI.
   - BBs may blunt the tachycardic response to hypoglycemia in diabetics, and worsening of hypoglycemia in diabetics on oral agents or insulin has been reported.

SUMMARY POINTS

- Anti-angina treatment has not been shown to modify the natural history of disease therefore the primary goal in the management of such patients is symptom control and amelioration of quality of life.
- Classical anti-angina drug therapy is based on treatment with nitrates, beta blockers, and calcium channel blockers.
- A combination therapy is often necessary and even when this strategy is adopted, a complete symptom relief may often not be the case.
- New anti-anginal drugs such as ivabradine, nicorandil, ranolazine and trimetazidine have therefore been developed.
- These drugs seem to be at least as beneficial as the classical ones.
- However, their beneficial effects are not due to hemodynamic effects, and therefore can be additive when used in combination with hemodynamic agents.
- A tailored therapy with a various combinations of the whole armamentarium (hemodynamic agents and 'non conventional therapy') appears therefore the best treatment strategy for angina patients.

DEFINITIONS

Ischemic heart disease (IHD): refers to one of these clinical syndromes: acute myocardial infarction, stable angina, unstable angina and cardiac death.

Coronary artery disease (CAD): refers to atherosclerotic disease of epicardial coronary arteries, the major cause of IHD. However, in clinical practice, these two entities are commonly interchanged, so that CAD is used instead of IHD to describe patients suffering one the previously mentioned clinical syndromes.

Preload: is the pressure that stretches the left ventricle before contraction and is mainly dependent on venous return. A greater preload is associated with a greater intraventricular pressure. This results in increased wall tension and myocardial oxygen consumption.

Afterload: is the obstacle (as measured by tension) towards which the heart expels the blood. The higher the afterload, the higher the energy requirements and oxygen consumption.

Optimal medical therapy (OMT): refers to the maximally tolerated drug therapy for symptomatic control with a various combination among BBs, nitrates and CCBs plus antiplatelet agents, ACE-I and statins as appropriate.

Hemodynamic agents: these drugs lower myocardial oxygen requirements reducing hemodynamic parameters (preload and/or afterload). This translates in lower heart rate and/or pressure and as such their effects are clinically 'more visible'.

Non hemodynamic agents: except for ivabradine which lowers heart rate, their action mechanism is different in that they optimize energy production capacity of the heart without altering hemodynamic parameters and therefore overall performance capacity.

Cardioselective BBs: preferentially inhibit β1 receptors that are principally found in the myocardium. Adverse effects associated with β2-adrenergic receptor antagonist activity (bronchospasm, peripheral vasoconstriction, alteration of glucose and lipid metabolism) are less common with β1-selective agents. However receptor selectivity diminishes at higher doses.

Non-cardioselective BBs: also inhibit β2 receptor sites, which are found in smooth muscle in the lungs, blood vessels, and other organs. Carvedilol and labetalol also have α-adrenergic receptor blocking properties. While α-antagonist properties are usually beneficial (i.e., patients with hypertension), β2 antagonist properties are often associated with worsened clinical status of patients with co-morbidities such as chronic obstructive pulmonary disease and peripheral artery disease.

I_f *current:* I_f current is part of the ionic currents that influence spontaneous diastolic depolarisation of the sinoatrial node, in this way determining the pacemaker activity. The 'I' stands for inward, while the 'f' denotes 'funny', so called because it had unusual properties compared with other current systems known at the time of its discovery. The I_f current is carried by both sodium and potassium ions across the sarcolemma; it is inward at voltages in the diastolic range, is activated on hyperpolarization (within the diastolic range of voltages regularly observed in cardiac pacemaker tissue) and is characterized by unusually low single-channel conductance and slow activation kinetics.

LIST OF ABBREVIATIONS

ACE-I	:	angiotensin converting enzyme inhibitor
ACS	:	acute coronary syndrome
ATP	:	adenosine triphosphate
BBs	:	beta blockers
BP	:	blood pressure
CABG	:	coronary artery bypass grafting
CAD	:	coronary artery disease
CCBs	:	calcium channel blockers
CIDs	:	clinically important differences
HF	:	heart failure
IHD	:	ischemic heart disease
MI	:	myocardial infarction
NO	:	nitric oxide
OMT	:	optimal medical therapy
PCI	:	percutaneous coronary intervention
SAN	:	sinoatrial node
SAQ	:	Seattle angina questionnaire

REFERENCES CITED

Boden, W.E., R.A. O'Rourke, K.K. Teo, P.M. Hartigan, D.J. Maron, W.J. Kostuk, M. Knudtson, M. Dada, P. Casperson, C.L. Harris, B.R. Chaitman, L. Shaw, G. Gosselin, S. Nawaz, L.M. Title, G. Gau, A.S. Blaustein, D.C. Booth, E.R. Bates, J.A. Spertus, D.S. Berman, G.B. Mancini and W.S. Weintraub. 2007. Optimal medical therapy with or without PCI for stable coronary disease. N Engl J Med 356(15): 1503–1516.

Chaitman, B.R. 2006. Ranolazine for the treatment of chronic angina and potential use in other cardiovascular conditions. Circulation 113(20): 2462–2472.

Effect of nicorandil on coronary events in patients with stable angina: the Impact Of Nicorandil in Angina (IONA) randomised trial. 2002. *Lancet* 359(9314): 1269–1275.

Elder, D.H., M. Pauriah, C.C. Lang, J. Shand, I.B. Menown, B.D. Sin, S. Gupta, S.G. Duckett, W. Foster, D. Zachariah and P.R. Kalra. Is there a Failure to Optimize theRapy in anGina pEcToris (FORGET) study? QJM 103(5): 305–310.

Fraker, T.D. Jr., S.D. Fihn, R.J. Gibbons, J. Abrams, K. Chatterjee, J. Daley, P.C. Deedwania, J.S. Douglas, T.B. Ferguson Jr., J.M. Gardin, R.A. O'Rourke, S.V. Williams, S.C. Smith Jr., A.K. Jacobs, C.D. Adams, J.L. Anderson, C.E. Buller, M.A. Creager, S.M. Ettinger, J.L. Halperin, S.A. Hunt, H.M. Krumholz, F.G. Kushner, B.W. Lytle, R. Nishimura, R.L. Page, B. Riegel, L.G. Tarkington and C.W. Yancy. 2007. chronic angina focused update of the ACC/AHA 2002 Guidelines for the management of patients with chronic stable angina: a report of the American College of Cardiology/American Heart Association Task Force on Practice Guidelines Writing Group to develop the focused update of the 2002 Guidelines for the management of patients with chronic stable angina. Circulation 116(23): 2762–2772.

Gibbons, R.J., J. Abrams, K. Chatterjee, J. Daley, P.C. Deedwania, J.S. Douglas, T.B. Ferguson Jr., S.D. Fihn, T.D. Fraker Jr., J.M. Gardin, R.A. O'Rourke, R.C. Pasternak and S.V. Williams. 2003. ACC/AHA 2002 guideline update for the management of patients with chronic stable angina—summary article: a report of the American College of Cardiology/American Heart Association Task Force on practice guidelines (Committee on the Management of Patients With Chronic Stable Angina). J Am Coll Cardiol 41(1): 159–168.

Grant, R.H., P. Keelan, R.J. Kernohan, J.C. Leonard, L. Nancekievill and K. Sinclair. 1966. Multicenter trial of propranolol in angina pectoris. Am J Cardiol 18(3): 361–365.

Holubkov, R., W.K. Laskey, A. Haviland, J.C. Slater, M.G. Bourassa, H.A. Vlachos, H.A. Cohen, D.O. Williams, S.F. Kelsey and K.M. Detre. 2002. Angina 1 year after percutaneous coronary intervention: a report from the NHLBI Dynamic Registry. Am Heart J 144(5): 826–833.

Hueb, W., P.R. Soares, B.J. Gersh, L.A. Cesar, P.L. Luz, L.B. Puig, E.M. Martinez, S.A. Oliveira and J.A. Ramires. 2004. The medicine, angioplasty, or surgery study (MASS-II): a randomized, controlled clinical trial of three therapeutic strategies for multivessel coronary artery disease: one-year results. J Am Coll Cardiol 43(10): 1743–1751.

Jahangir, A., A. Terzic and W.K. Shen. 2001. Potassium channel openers: therapeutic potential in cardiology and medicine. Expert Opin Pharmacother 2(12): 1995–2010.

Kardas, P. 2007. Compliance, clinical outcome, and quality of life of patients with stable angina pectoris receiving once-daily betaxolol versus twice daily metoprolol: a randomized controlled trial. Vasc Health Risk Manag 3(2): 235–242.

Koester, R., J. Kaehler, H. Ebelt, G. Soeffker, K. Werdan and T. Meinertz. Ivabradine in combination with beta-blocker therapy for the treatment of stable angina pectoris in every day clinical practice. Clin Res Cardiol 99(10): 665–672.

Markham, A., G.L. Plosker and K.L. Goa. 2000. Nicorandil. An updated review of its use in ischaemic heart disease with emphasis on its cardioprotective effects. Drugs 60(4): 955–974.

McCormack, J.G., R.L. Barr, A.A. Wolff and G.D. Lopaschuk. 1996. Ranolazine stimulates glucose oxidation in normoxic, ischemic, and reperfused ischemic rat hearts. Circulation 93(1): 135–142.

Otsuka, T., C. Ibuki, T. Suzuki, K. Ishii, H. Yoshida, E. Kodani, Y. Kusama, H. Atarashi, H. Kishida, T. Takano and K. Mizuno. 2008. Administration of the

Rho-kinase inhibitor, fasudil, following nitroglycerin additionally dilates the site of coronary spasm in patients with vasospastic angina. Coron Artery Dis 19(2): 105–110.

Pepine, C.J., J. Abrams, R.G. Marks, J.J. Morris, S.S. Scheidt and E. Handberg. 1994. Characteristics of a contemporary population with angina pectoris. TIDES Investigators. Am J Cardiol 74(3): 226–231.

Pepine, C.J., J.L. Rouleau, K. Annis, A. Ducharme, P. Ma, J. Lenis, R. Davies, U. Thadani, B. Chaitman, H.E. Haber, S.B. Freedman, M.L. Pressler and B. Pitt. 2003. Effects of angiotensin-converting enzyme inhibition on transient ischemia: the Quinapril Anti-Ischemia and Symptoms of Angina Reduction (QUASAR) trial. J Am Coll Cardiol 42(12): 2049–2059.

Rousseau, M.F., H. Pouleur, G. Cocco and A.A. Wolff. 2005. Comparative efficacy of ranolazine versus atenolol for chronic angina pectoris. Am J Cardiol 95(3): 311–316.

Rumsfeld, J.S., D.J. Magid, M.E. Plomondon, A.E. Sales, G.K. Grunwald, N.R. Every and J.A. Spertus. 2003. History of depression, angina, and quality of life after acute coronary syndromes. Am Heart J 145(3): 493–499.

Ruzyllo W., M. Tendera, I. Ford and K.M. Fox. 2007. Antianginal efficacy and safety of ivabradine compared with amlodipine in patients with stable effort angina pectoris: a 3-month randomised, double-blind, multicentre, noninferiority trial. Drugs 67(3): 393–405.

Thadani, U. 1997. Nitrate tolerance, rebound, and their clinical relevance in stable angina pectoris, unstable angina, and heart failure. Cardiovasc Drugs Ther 10(6): 735–742.

Vicari, R.M., B. Chaitman, D. Keefe, W.B. Smith, S.G. Chrysant, M.J. Tonkon, N. Bittar, R.J. Weiss, H. Morales-Ballejo and U. Thadani. 2005. Efficacy and safety of fasudil in patients with stable angina: a double-blind, placebo-controlled, phase 2 trial. J Am Coll Cardiol 46(10): 1803–1811.

Wasserstrom, J.A., R. Sharma, M.J. O'Toole, J. Zheng, J.E. Kelly, J. Shryock, L. Belardinelli and G.L. Aistrup. 2009. Ranolazine antagonizes the effects of increased late sodium current on intracellular calcium cycling in rat isolated intact heart. J Pharmacol Exp Ther 331(2): 382–391.

Wei, J., T. Wu, Q. Yang, M. Chen, J. Ni and D. Huang. Nitrates for stable angina: A systematic review and meta-analysis of randomized clinical trials. Int J Cardiol 146(1): 4–12.

Current Clinical Application of Direct Thrombin Inhibitors in Angina Pectoris

Bernardo Cortese[1,]* and *Marco Centola*[2]

ABSTRACT

Thrombin is a key mediator of the coagulation cascade and also exerts a crucial role in promoting platelet aggregation. Since decades, drugs that block thrombin activity have been widely studied for the treatment of stable and unstable angina pectoris. First were the heparins (both unfractionated and with light molecular weight), however this class of drugs showed some drawbacks that limit their use, especially in those patients that are at higher risk of thrombotic and hemorragic complications. More recently, direct thrombin inhibitors have been widely studied for many clinical conditions, among others the complications of heparin induced thrombocytopenia, prophylaxis and treatment of deep venous thrombosis, stable and unstable angina pectoris, myocardial infarction, percutaneous coronary interventions, atrial fibrillation. After a pharmacokinetic and pharmacodynamics appraisal of this class of drugs, we will focus on clinical applications of direct thrombin inhibitors for the treatment of patients suffering

---

[1]Interventional Cardiology, Cliniche Humanitas Gavazzeni, Bergamo, Italy; Email: bcortese@gmail.com
[2]Interventional Cardiology, Azienda Ospedaliera San Paolo, Milan, Italy; Email: marco.centola@unimib.it
*Corresponding author

List of abbreviations after the text.

angina pectoris, analyzing the improvements on heparins but also the disadvantages. We will also focus on the bleeding issue, that recently became a major factor for the outcome of cardiologic patients. Newer modalities of infusion of bivalirudin will also be discussed. DTIs drawbacks are described, as well as their advantages over heparins. A review of all available data regarding clinical trials on direct thrombin inhibitors are also reported.

INTRODUCTION

Thrombin, a serine protease, is an enzyme that results from the cleavage of activated factor X on its precursor (prothrombin, derived by the liver). Thrombin is a key mediator of the coagulation cascade: once both the extrinsic and intrinsic pathways of coagulation have activated enough Factor X, prothrombin is converted to thrombin, the final activator of fibrinogen to fibrin. Moreover, thrombin is a very important mediator of platelet aggregation via the PAR-1 receptors on platelet surface. By activating factor XIII, it leads to the formation of cross-linked bonds among fibrin molecules, improving clot stabilization (Topol 2001, Coughlin 2000).

The molecule of thrombin has different recognition sites for the adhesion of molecules: an active (catalytic) site, exosite 1 and exosite 2. Fibrinogen binds exosite 1, whereas the complex heparin-ATIII binds exosite 2, and in case of formation of a complex fibrin-heparin-thrombin, it occupies both exosites. Direct thrombin inhibitors (DTI) are a class of drugs counted among the anticoagulants, that elicit their activity by inhibiting thrombin without the intermediation of cofactors (the ATIII) (Fig. 8.1) (Lane et al. 2005).

Due to the central role of thrombin in thrombus formation with implications in both the coagulation cascade and platelet aggregation, drugs that block its activity have been widely studied. In animal models, DTIs have shown to inhibit platelet and fibrinogen deposition on fresh thrombus, to reduce the increase in thrombus size and to dissolve existing thrombus (Weitz et al. 1990, Meyer et al. 1998, Meyer et al. 1994, Kam et al. 2005).

Current treatment with heparins carries many limitations at least in part overcome by DTIs (Table 8.1).

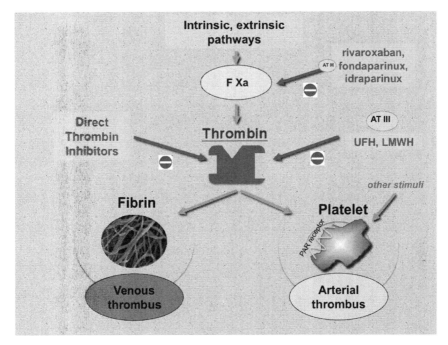

Figure 8.1 Thrombin central role for the formation of clots, with its antagonists.

The anticoagulant activity of unfractionated heparin (UFH) and low molecular weight heparins (LMWH) are mediated by the interaction with antithrombin III. Heparins generate a ternary heparin–thrombin–antithrombin complex, leading to thrombin inhibition. The activity of DTIs is independent of the presence of antithrombin and is related to the direct interaction of these drugs with the thrombin molecule. Unpublished.

Color image of this figure appears in the color plate section at the end of the book.

DTIs have been widely studied for many clinical conditions: complications of heparin-induced thrombocytopenia (HIT) (Shantsila et al. 2009), prophylaxis and treatment of deep venous thrombosis (Eriksson et al. 2011), stable angina (Lincoff et al. 2003), acute coronary syndromes (ACS) (Stone et al. 2006), percutaneous coronary interventions (PCI) (Lincoff et al. 2003), non-valvular atrial fibrillation (Wallentin et al. 2010). In this chapter we will thoroughly analyze those DTIs that have been studied in patients suffering angina pectoris.

Table 8.1 Differences between indirect and direct thrombin inhibitors.

Heparins' drawbacks	Clinical consequence	What a DTI may add
Indirect thrombin inhibitory activity (require ATIII)	Reduced efficacy in case of deficit of A III (mostly for UFH)	Direct thrombin inhibition
Platelet aggregation promotion (GPIIb/IIIa receptor)	Reduced efficacy in case of absence of antiplatelet agents	Mild antiplatelet effect, inhibiting thrombin-PAR interaction
Nonlinear pharmacokinetics (need of aPTT)	Poor bioavailability, variable effect (for UFH)	Linear pharmacokinetics; predictable and dose-related anticoagulation
Binding and dependence on plasma proteins	Poor bioavailability, variable effect	Non-binding to plasma proteins (bivalirudin)
Lack of inhibition of platelet and fibrin-bound thrombin	Reduced efficacy	Inhibition of platelet and fibrin-bound thrombin
Immune-mediated risk of HIT	HIT (mostly for UFH)	Immunologically inert; no risk of thrombocytopenia
Vascular permeability	Increase of tissue edema and ischemia (UFH)	DTIs do not increase vascular permeability

Main drawbacks of heparins, their mechanism and clinical consequences, and what a DTI may add. DTI=direct thrombin inhibitor; ATIII=antithrombin III; UFH=unfractionated heparin; GPIIb/IIIa=glycoprotein IIb/IIIa; ACT=activated clotting time; aPTT=activated partial thromboplastin time; HIT=heparin-induced-thrombocytopenia. Unpublished.

PHARMACOKINETICS AND PHARMACODYNAMICS OF DIRECT THROMBIN INHIBITORS

The natural anticoagulant of leeches is hirudin, now available through recombinant DNA technology (desirudin/lepirudin): it is a 65-aminoacid protein that binds thrombin very tightly, with high specificity and low dissociation rate, at both the active site and exosite 1. One of the most important peculiarities common to all DTIs, is that these drugs do not activate platelets, therefore cannot cause HIT (Hirsh 2001). Another key point is that all DTIs may also block clot-bound and fibrin-bound thrombin.

Low molecular weight DTIs have also been developed, divided into two classes: the non-covalent (argatroban, napsagatran, inogatran, melagatran, dabigatran) and the reversible-covalent (efegatran) DTIs. Argatroban, melagatran, efegatran, inogatran are also classified as univalent DTIs because they only bind thrombin on the catalytic site, whereas hirudin and bivalirudin bind thrombin both at the catalytic site and exosite 1 (Fig. 8.2).

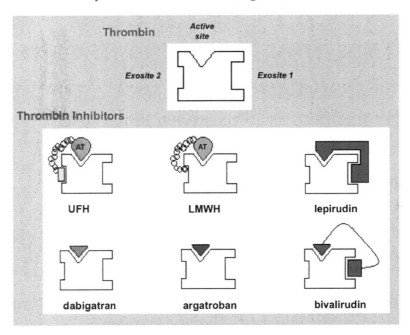

Figure 8.2 Thrombin molecule and interaction with mediators.

Legend: there are three domains on the thrombin molecule: an active site and two opposite located exosites. UFH binds exosite 2 and antithrombin III, which binds the enzyme active site. LMWH lack the longer chains of UFH that bind exosite 2. Lepirudin binds directly both the active enzymatic site and exosite 1 of thrombin, forming a slowly reversible complex. Bivalirudin binds thrombin in a fast, reversible fashion. Argatroban and dabigatran are small molecules that bind reversibly to the active enzymatic site of thrombin. Exosite 1=fibrin/fibrinogen binding site; Exosite 2=heparin-binding site; UFH=unfractionated heparin; LMWHs=low-molecular-weight-heparins; AT=antithrombin; DTI=direct trhombin inhibitor. Unpublished.

Color image of this figure appears in the color plate section at the end of the book.

Bivalirudin is a semi-synthetic peptide of 20 aminoacids, and is currently the most utilized DTI in catheterization laboratories. It has a bivalent bond to the two same active sites of hirudin, but different from its precursor exerts only a transient inhibition on thrombin due to a reversible bond caused by rapid cleavage by thrombin; it follows rapid plasma clearance and low renal or hepatic metabolism. The estimated mean half life of bivalirudin is 25 min. One of the most important peculiarities of bivalirudin is its mild antiplatelet effect exerted with the inhibition of the interaction of thrombin and thrombin PAR receptor (Bates and Weitz 1998). Another key point is that its reversible effect along with the short half life allows fast dissociation with thrombin and rapid hemostasis after discontinuation.

Ximelagatran and argatroban are synthetic derivates. Ximelagatran, an oral prodrug, is metabolized to the active form melagatran after gastrointestinal absorption. On the other hand, melagatran can be administered subcutaneously, and has a half life of 2-3 hr (ximelagatran has 3–5 hr). Argatroban is given intravenously without a bolus and reaches its steady state in plasma in 8-12 hr. Its plasma half life is 45 min. Different from all the other DTIs, this drug is biliary excreted, therefore in patients with severe hepatic dysfunction a dose adjustment is required. Due to this metabolism, it has shown to increase the INR when given with warfarin (Hirsh et al. 2007).

Dabigatran etexilate is an orally given prodrug rapidly converted to dabigatran by enterocytes and in the liver via a serum esterase independent of cytochrome P-450. It exerts rapid onset of action with reversible effect. Dabigatran half life is 12–17 hr. This drug is renally excreted in its greatest part, only a minor part being conjugated with glucuronic acid and biliary excreted (Hirsh et al. 2007).

All commercially available DTIs, with the exception of argatroban, have a predominant renal clearance and could accumulate in patients with impaired kidney function. Therefore, it should be kept in mind that in such cases DTIs plasma half life may significantly be prolonged. Data on long term use of ximelagatran and melagatran indicate that liver enzymes elevation can occur in more than 8% of patients after 6 mon of treatment. In the majority of cases, the hepatic dysfunction is asymptomatic and reversible, even if the treatment is not interrupted. DTIs do not have relevant interaction with food, aspirin or drugs that are metabolized via the cytochrome

P450. Of note, fatal anaphylaxis has been described with lepirudin, particularly in patients previously treated with this drug. In contrast, argatroban and bivalirudin do not have immunogenic properties (Hirsh et al. 2007).

There are two methods for monitoring activity of DTIs, the ecarin clotting time (ECT) and the activated clotting time (ACT), the latter being the most widely used. In addition, there is no antidote for radiply reversing DTI anticoagulant properties.

Table 8.2 shows current indications to the use of DTIs. In the following paragraphs, we will analyze modern DTI use for the treatment of angina pectoris in the ward and in the catheterization laboratory.

Table 8.2 Current indications of direct thrombin inhibitors.

Drug	Administration	Current clinical indications for DTIs
Desirudin	sc	Venous thromboembolism prevention in orthopedic elective surgery (USA, Europe)
Lepirudin	iv, sc	Thrombosis treatment and prevention in patients with history of HIT (USA, Europe)
Bivalirudin	iv	Unstable angina/myocardial infarction with planned revascularization; PCI (USA, Europe); thrombosis treatment and prevention in patients with history of HIT undergoing PCI (USA)
Argatroban	iv	Thrombosis treatment and prevention in patients with history of HIT undergoing or not PCI (USA)
Dabigatran etexilate	os	Venous thromboembolism prevention in orthopedic surgery (Europe, Canada); stroke prevention in atrial fibrillation (USA)

All commercially available DTIs are reported in this table, with the administration and current clinical indication. DTI=direct thrombin inhibitor; sc=subcutaneous; iv=intravenous; HIT=heparin-induced-thrombocytopenia; PCI=percutaneous coronary intervention. Unpublished.

PRACTICE AND PROCEDURES: CLINICAL APPLICATIONS OF DIRECT THROMBIN INHIBITORS IN ANGINA PECTORIS

Patients not Undergoing Coronary Intervention

Hirudin

The GUSTO IIb trial enrolled 12142 patients with unstable angina and myocardial infarction that were randomized to 72-hr infusion of heparin and hirudin. The primary study endpoint, 30-d death and myocardial infarction (MI), was lower in the hirudin group (6.8 vs 9.6%, p = 0.06), mostly due to lower myocardial infarction (MI) rates, that were significantly lower in this group. If life threatening or other major bleedings were not influenced by treatment allocation, hirudin-treated patients experienced an higher risk of minor bleedings (8.8% vs 7.7%, p = 0.03) (GUSTO IIb investigators 1996). On the other hand, the OASIS-2 trial enrolled 10141 patients with unstable angina and non-ST elevated MI randomized to hirudin and UFH (72 hr infusion in both groups). Cardiovascular death and MI after 7 d, primary study endpoint, were not reduced by hirudin, but composite secondary endpoint of death, MI or refractory angina was significantly lower (5.6% vs 6.7%, p = 0.0125). Patients undergoing early PCI and treated with hirudin experienced a significant reduction in 30-d death and myocardial infarction (OASIS-2 Investigators 1999).

However, despite the more prominent anti-ischemic effect shown in these trials, safety concerns were raised by the occurrence of increased bleeding in patients treated with hirudin, mostly due to its narrow therapeutic window. In the OASIS-2 trial the drug had to be interrupted for hemorrhages or adverse events in 2.8% vs 1.3% in the heparin group (p < 0.001).

Further experimentations with hirudin were abandoned, due to the higher risk of hemorrhages and the lack of a clear benefit over UFH.

Argatroban

In a Phase I clinical trial, Gold et al studied the antithrombotic and clinical effects of argatroban in 43 patients with unstable angina pectoris. The intravenous infusion of the drug for 4 hr was monitored by sequential measurements of coagulation times and indexes of thrombin activity in vivo followed by a 24-hr clinical observation period. The tested drug seemed to prolong the aPTT and inhibit thrombin activity toward fibrinogen (decrease of fibrinopeptide A), but in vivo thrombin and thrombin-ATIII complex formation was not suppressed. Moreover, the cessation of infusion was associated with rebound thrombin generation with an early dose-related recurrence of unstable angina. The authors concluded that although the mechanism of this clinical and biochemical rebound phenomenon remains to be determined, its implication for argatroban use in ACS might be significant (Gold et al. 1993).

Efegatran

In 1999, Klootwijk enrolled 432 patients with UA to efegatran and UFH to evaluate the ability of this DTI to suppress myocardial ischemia. Five dose levels of efegatran were studied sequentially over 48 hr. Administration of efegatran at dosages >0.63 mg/kg/h warranted an anti-ischemic activity similar to that of UFH, and bleeding did not increase. The level of thrombin inhibition by efegatran, as measured by aPTT, appeared to be more stable than with UFH. The authors concluded that despite these results, no clear advantage over UFH was demonstrated (Klootwijk et al. 1999).

Inogatran

Inogatran was tested in the TRIM study, where 1209 patients with UA or non-Q wave MI were randomly assigned to inogatran or UFH repetitive boluses and subsequent 72-hr infusion. UFH-treated patients experienced fewer episodes of ischemia on Holter monitoring, but combined primary endpoint incidence (death, recurrence of angina pectoris and MI) was similar in the study groups. Major bleeding was low and not dissimilar. This study, too, was not able to demonstrate any clinical benefit of inogatran (TRIM Study Group 1997).

Ximelagatran

In the ESTEEM trial various dosages of the oral DTI ximelagatran and placebo were compared in patients with recent MI. The study drug reduced the incidence of all-cause mortality, non fatal MI and severe recurrent ischemia at six-mon follow up. However, the investigators reported a slight (although not significant) increase of bleeding, and a significant elevation of liver transaminases, which was dose-related. It should be noted that patients were not pretreated with a thienopyridine (Wallentin et al. 2003).

A meta-analysis of the Direct Thrombin Inhibitor Trialist Collaborative Group included 11 randomized trials and 35970 patients with angina pectoris or ACS assigned to up to 7 d treatment with DTIs or UFH. Compared with UFH, the DTIs were associated with a lower risk of death or MI at the end of treatment and at 30-d follow up. This was due primarily to a reduction in MIs with no apparent effect on deaths. The benefit was evident in all categories of trials, and it was demonstrated with both hirudin and bivalirudin. The positive effect of DTIs might be explained by the purely anticoagulant action but also by the mild antiplatelet activity, the two key elements of intracoronary thrombus formation, whereas other drugs only exert antiplatelet activity and require the addition of an anticoagulant such as UFH. However, subgroup analyses showed that hirudin was associated with an increased risk of major bleeding, particularly in patients suffering an ACS. By contrast, bivalirudin was associated with a 50% reduction in the major bleeding risk, mainly in patients undergoing PCI (DTI Trialists' Collaborative Group 2002).

Bivalirudin

The interesting pharmacokinetic and parmacodynamic properties of bivalirudin rendered it suitable for experimentations in various clinical settings.

The Bivalirudin Angioplasty Trial was the first study to provide data on the safety and efficacy of bivalirudin infusion. In this trial, 4312 patients with unstable or postinfarction angina were randomized to bivalirudin or UFH. The primary composite endpoint of death, MI, or revascularization at 7 d occurred significantly less in the bivalirudin group (6.2 vs. 7.9%, P=0.039). Differences persisted at 90 d. In addition to reduced adverse ischemic events, bivalirudin therapy was associated with significantly fewer bleedings (3.5 vs. 9.3%, P<0.001) (Bittl et al. 1995).

A different dosage of bivalirudin was however studied in the ACUITY trial, that enrolled 13819 patients with moderate or high-risk ACS without ST-segment elevation (41% of patients suffered unstable angina). Patients were randomized to heparin+GP IIb/IIIa inhibitor (GPI), bivalirudin plus GPI, or bivalirudin with provisional GPI. An additional IV bolus of bivalirudin was administered in those patients that underwent PCI. For the primary 30-d endpoint, a composite net clinical adverse outcome of ischemia endpoints (death, MI, or ischemia-driven revascularization) + protocol major hemorrhages, bivalirudin and provisional GPI (used in 10% of patients) was found superior to the other two treatments, due to lower per protocol bleeding complications. Also TIMI major and minor bleedings were lowered by bivalirudin alone treatment. Interestingly, ischemic endpoints were similar between the three treatment arms (Stone et al. 2006).

Patients with Angina Pectoris Undergoing Coronary Intervention

In the HELVETICA trial 1141 patients with unstable angina undergoing PCI were randomized to heparin bolus and infusion vs. desirudin bolus and infusion. Ninety six-h follow up showed reduced clinical ischemic events in the desirudin patients. No significant differences were shown at 30-wk clinical follow up. However, most unstable patients seemed to benefit from hirudin treatment. No differences in major and minor bleedings were recorded (Herrman et al. 1995).

In 2001 Lewis et al enrolled 91 patients with HIT with indication to PCI and treatment with intravenous argatroban. Outcomes on intravenous argatroban were compared with historical control patients treated with UFH. In this setting, argatroban was found safe because no patient experienced death at prespecified 24-hr follow up, whereas four patients experienced MI and four unplanned revascularization; at least one MACE occurred in seven patients overall. Recently, Jang et al evaluated the efficacy of argatroban+GPI in patients undergoing PCI for ACS. The primary endpoint of the study (death, MI or need for repeat revascularization at 30 d post stenting) occurred in 3% of patients. No further studies with longer follow up are currently available (Lewis et al. 2001).

The reduced recurrence of MI in the HERO-2 study, that enrolled patients with MI, stimulated the interest of bivalirudin in the cath lab. In the REPLACE-2 trial 6010 patients undergoing PCI for stable or unstable angina pectoris or MI were randomized to bivalirudin or UFH+GPI. Bivalirudin was administered as a cath lab bolus of 0.75 mg/Kg, then at an infusion rate of 1.75 mg/Kg/h for the duration of the procedure. The composite net clinical adverse outcome at 30 d, as well as the composite ischemic endpoints were similar in the two groups. This trial confirmed that both major and minor bleedings were reduced by bivalirudin treatment (Lincoff et al. 2003).

The worst composite ischemic endpoint in those patients that were not preloaded with clopidogrel in the ACUITY trial, and an excess of acute stent thromboses in the HORIZONS-AMI trial in patients undergoing primary PCI (Mehran et al. 2009), lift a veil of shadow on bivalirudin and suggested to our group to prolong its infusion for 4-hr after the intervention. The rationale was to protect the myocardium just when the thrombotic risk was higher, namely in the first hours after the intervention. Therefore, we conducted the PROBI VIRI study, a randomized clinical trial addressing the efficacy of a prolonged bivalirudin infusion after complex PCI. One-hundred-seventy-eight patients with angina pectoris and a clinical indication to multivessel or complex PCI were enrolled to standard or prolonged bivalirudin infusion. Study primary endpoint, procedural-related MI, was found significantly lower in the prolonged infusion arm without increasing the risk of bleedings, thus suggesting a protective role of this drug from acute stent thrombosis (Fig. 8.3) (Cortese et al. 2009).

Further studies may be needed to assess the correct timing of antiplatelet and DTI administration in patients with angina or ACS. In this regard, a recent post-hoc analysis of the ACUITY trial showed that, as long as clopidogrel is administered before or within 30 min after PCI, bivalirudin alone warrants similar outcome to heparin+GPI in suppressing 30-d ischemic events, with significantly fewer bleedings. If clopidogrel is given later, bivalirudin may be associated with worse ischemic outcomes. In this regard, the prolonged infusion of bivalirudin already cited might help in suppressing this excess of ischemic events, as previously shown (Cortese 2009).

End points of study			
Variable	Standard Bivalirudin (n = 90)	Prolonged Bivalirudin (n = 88)	p Value
In hospital			
CK-MB ≥3 times ULN (primary end point)	15 (17%)	6 (7%)	0.041
Peak postprocedural CK-MB (ng/ml), mean ± SD	10.4 ± 26.4	4.9 ± 11.1	0.044
Major bleedings	1 (1%)	1 (1%)	0.87
Minor bleedings	3 (3%)	3 (3%)	0.96
30-d follow-up			
MACEs	3 (3%)	1 (1%)	0.43
Death	1 (1%)	0 (0%)	0.37
Q-wave myocardial infarction	2 (2%)	1 (1%)	0.70
Target vessel revascularization	1 (1%)	0 (0%)	0.37
6-mo follow-up			
MACEs	15 (17%)	9 (10%)	0.18
Death	2 (2%)	1 (1%)	0.70
Q-wave myocardial infarction	4 (4%)	3 (3%)	0.76
Target vessel revascularization	11 (12%)	7 (8%)	0.33
Definite/probable stent thrombosis*	0 (0%)	0 (0%)	1.00

* As defined by the Academic Research Consortium.

Figure 8.3 Results of the Probi Viri study.

The main results of the Probi Viri study are reported here, that show a reduced risk of periprocedural myocardial infarction after a prolonged infusion of bivalirudin in patients with angina pectoris undergoing high-risk coronary intervention. No differences were discovered for clinical endpoints. MACEs=major adverse clinical events. Permission obtained by Elsevier editor.

THE BLEEDING ISSUE WITH DIRECT THROMBIN INHIBITORS

In the last 10 yr the use of PCI became widespread and conversely the clinical indications to percutaneous interventions became wider. Concomitantly, the use of potent antithrombotic drugs has lowered ischemic complication rates and warranted higher procedural success. However, that happened at the cost of higher bleedings. For these results among others, bleedings have reached the 'entity' of complications, and are nowadays often considered to impact on survival and quality of life of patients at the same level of purely ischemic complications. Many studies have also unequivocally demonstrated that hemorrhages resulted with impaired survival.

Table 8.3 DTIs tested in randomized clinical trials.

Trial name	DTI/control	Clinical setting (n° patients)	Primary endpoint	Ischemic endpoints	Bleeding endpoints
GUSTO IIb	hirudin/UFH	AMI (4131), UA (8011)	composite end-point death and MI	14% reduction composite end-point death and MI	equivalent
HIT IV	hirudin/UFH	AMI (1200)	angiographic TIMI 3 flow 90 min	no mortality effect; better initial ST resolution	equivalent
OASIS-2	hirudin/UFH	UA (10141)	combined end point death and MI	12% decrease in MI, death and reintervention	equivalent
HELVETICA	hirudin/UFH	PCI (1141)	all cardiac events	decreased reintervention	equivalent
HERO-2	bivalirudin/UFH for 48 hr	MI with or without ST elevation (17073)	death at 30 d	no difference in mortality; reduction of reinfarction within 96 hr by 30%	moderate/mild bleedings higher
ACUITY	UFH or enoxaparin plus GP IIb/IIIa inhibitor, bivalirudin plus GP IIb/IIIa inhibitor, bivalirudin monotherapy	ACS (13819)	death, MI, unplanned revascularization and mortality at 1 yr	equivalent	lower bleedings
HORIZONS AMI	bivalirudin/UFH plus GP IIb/IIIa inhibitor	STEMI (3602)	net adverse clinical events (NACEs), (combination of major bleedings and MACE)	lower cardiovascular mortality	lower bleedings
PROTECT TIMI-30	eptifibatide+UFH, eptifibatide+enoxaparin, or bivalirudin	non-ST elevation ACS+PCI (854)	coronary flow reserve	equivalent	equivalent

Current Clinical Application of Direct Thrombin Inhibitors in Angina Pectoris   147

REPLACE-2	bivalirudin/UFH plus GP IIb/IIIa for 12-18 hr	PCI (6010)	death, MI, urgent repeat revascularization, serious bleeding	equivalent	lower serious bleedings
PROBI VIRI	bivalirudin standard dose/ bivalirudin prolonged infusion	angina pectoris+PCI (178)	procedural-related necrosis	reduced risk of mionecrosis with prolonged infusion; no differences for MACE	equivalent
MINT	argatroban vs UFH	AMI (125) treated with tPA	TIMI grade 3 flow at 90 min, rate of death and MI at 30 d	18% increase in TIMI 3 flow, decreased death and MI	lower major bleedings
ARGAMI-2	argatroban vs UFH	AMI (1200) treated with tPA and SK	mortality, MI and angiographic patency at 30 d	equivalent	lower bleedings
TRIM	inogatran or UFH bolus doses	ACS (1209)	death and MI	equivalent	equivalent
Klootwijk P (1999)	efegatran versus UFH	UA (430)	angina, MI, coronary intervention, death	equivalent	equivalent
ESTEEM	ximelagatran/placebo for 6 mon	AMI with or without ST elevation (1883)	death, non fatal MI, severe ischemia	lower MACEs	lower serious bleedings
SPORTIF III	ximelagatran/warfarin for 17 mon	nonvalvular AF (3410)	all strokes and systemic embolism	equivalent	lower serious bleedings
SPORTIF V	ximelagatran/warfarin for 20 mon	nonvalvular AF (3922)	all strokes and systemic embolism	equivalent	lower serious bleedings
RE-LY	dabigatran/warfarin for more than 12 mon	AF (18113)	all strokes and systemic embolism	low dose equivalent/ high dose superior	lower bleedings

All available clinical trials with DTIs are in this table; we report actual dosage used and the primary efficacy and safety endpoint. DTI=direct thrombin inhibitor; AT III=antithrombin III; UFH=unfractionated heparin; AMI=acute myocardial infarction; STEMI=ST-elevation AMI; UA=unstable angina; sc=subcutaneous; PCI=percutaneous coronary intervention; MACE=major adverse clinical events; GPIIb/IIIa=glycoprotein IIb/IIIa; tPA=tissue plasminogen activator; AF=atrial fibrillation. Unpublished.

Moreover, the hemorrhage itself, and often also its treatment with blood transfusion and the need of suspending antithrombotic drugs are associated with increased ischemic complication rates (Doyle et al. 2009).

What is strange with DTIs, is that if the first studies with this class of molecules showed an impaired/inconsistent safety profile when the drug was compared to heparins, subsequent investigations with bivalirudin and newer dosages transformed, at least this drug, as the drug of choice in patients at medium and high bleeding risk with angina pectoris and ACS undergoing PCI. In fact, among others the ACUITY, HORIZONS-AMI, ISAR REACT-3 studies definitely gave this molecule the patent of safe drug, specifically indicated to achieve lower hemorrhages, both major and minor if compared with standard treatment, at the same time warranting comparable outcome regarding the ischemic endpoints.

CONCLUSIONS

Heparins are a mainstay among antithrombotic drugs. DTIs represent an evolution from heparins, and in the last 15 yr this class of drugs has been widely studied in many clinical conditions. However, some of these molecules failed to show a significant clinical improvement from the gold standard treatment, and sometimes appeared to have a worse safety profile.

The treatment of angina pectoris, both stable and unstable, is a relevant issue for cardiologists, and due to the lower risk of patients with this condition, if compared with those that experience an ACS, the antithrombotic drugs employed should maintain the same anti-ischemic effect of the gold standards. However, an improvement in terms of safety is advocated. Here bivalirudin has been widely studied in the intensive coronary unit and, especially, in the catheterization laboratory, and has shown to warrant a better safety profile reducing both major and minor bleedings. Newer modalities of administration like a prolonged post PCI infusion seem promising to fill the gap with more potent antithrombotic agents. On the other hand, newer oral antiplatelet drugs like prasugrel and ticagrelor might well combine with bivalirudin and further sperimentations will clarify if the standard safety profile of bivalirudin will be preserved.

KEY FACTS OF DTIS FOR ANGINA PECTORIS TREATMENT

- Angina pectoris is a chest pain, and usually reflects an atherosclerotic disease involving the coronary tree. It may precipatate a heart attack, therefore its rapid diagnosis and tempestive treatment is crucial. Since its first description from Heberden in 1768 in London, great improvements have been achieved. However, despite technical and scientific improvements, angina pectoris is still a challenge for modern cardiologists. Angina pectoris is classified as stable and unstable.
- Many drugs are involved for its treatment. Some (betablockers, calcium-channel blockers) lower oxygen consumption. Some others (e.g., nitrates) exert a vasodilatory effect. While others have an anticoagulant or antiplatelet effect. Among the anticoagulants heparins, exert an anti-FX and anti-FII effect.
- Thrombin (FII) is a key mediator of both the coagulation cascade and of platelet aggregation. Thrombin has three different domains for the adhesion of molecules: an active site and two opposite located exosites. DTIs interact directly with thrombin without the intermediation of cofactors. Argatroban, melagatran, efegatran, inogatran are univalent DTIs because they only bind thrombin on the active site, whereas hirudin and bivalirudin (known as bivalent DTIs) bind thrombin both at the active site and at the exosite 1.
- All DTIs have similar action and overcome the limitations of heparins. There are many applications in modern cardiology: atrial fibrillation, deep venous thrombosis, acute and subacute pulmonary embolysm, heparin induced thrombocytopenia. In patients suffering angina pectoris or acute coronary syndromes the consistent results of randomized trials and meta-analysis provide evidence of the superiority of DTIs. This is particularly true for bivalirudin. Other DTIs have inconsistent clinical evidence in this setting.
- There are some drawbacks of DTIs. Long term administration of melagatran, ximelagatran and argatroban has been associated with an elevation of the aminotransferase serum levels in some patients. Hirudin administration during clinical trials was associated with higher risk of bleeding compared to

heparin, immunogenicity, rebound ipercoagulability and strong dependence on renal function. Unlike the other DTIs, bivalirudin proved a good efficacy and an excellent safety profile in many different clinical settings. Bivalirudin is specifically indicated to achieve lower hemorrhages, compared with standard treatment, warranting comparable outcome in terms of ischemic endpoints.

SUMMARY

- Thrombin is a key mediator of coagulation cascade. It has three recognition sites for the adhesion of molecules. Heparins, a mainstay among antithrombotic drugs, indirectly inhibit thrombin through an antithrombin-dependent action. Heparins have some drawbacks. Among others, they increase the bleeding risk, may promote platelet aggregation, may cause heparin-induced thrombocytopenia and have poor bioavailability.
- Direct thrombin inhibitors (DTIs) are a novel class of drugs developed to overcome the limitations of heparins. There are many commercially available DTIs. Among others, one should mention hirudin, bivalirudin, argatroban and dabigatran. DTIs have been tested in various clinical settings, with variable efficacy and safety profiles. They have been widely studied for: complications of heparin-induced thrombocytopenia, prophylaxis and treatment of deep venous thrombosis, stable angina, acute coronary syndromes, percutaneous coronary interventions (PCI), non-valvular atrial fibrillation.
- In this chapter attention is focused on the treatment of angina pectoris. Patients should be divided into those that undergo and those that do not undergo PCI. The encouraging results observed in the trials of bivalirudin in patients with angina pectoris, and the consistent reduction in bleeding complications, make this agent an attractive alternative to UFH. In the Acuity trial it was shown to achieve improved net clinical outcome when compared to heparin+GP IIb/IIIa. Similar results have been achieved in patients with angina pectoris undergoing PCI.
- Some drawbacks of DTIs are still present, leading to the failure of few of them in clinical trials. Ximelagatran has been retrieved

from commerce due to supposed hepatic injury. Hirudin has an unfavorable pharmacokinetic profile, and has been abandoned due to a higher bleeding risk. Argatroban has not yet been studied in phase III trials. Efegatran and inogatran showed no particular benefit in this setting.
- Bivalirudin showed an unfavorable profile in terms of protection from myocardial infarction in those patients undergoing PCI and not pretreated with a thienopyridine. Moreover, it has been shown to have an increased risk of acute stent thrombosis. Newer modalities of bivalirudin administration like a prolonged post-PCI infusion seem promising to fill the gap with more potent antithrombotic agents. The Probi Viri study, discussed here, provided important data regarding this topic. Newer oral antiplatelet drugs might well combine with bivalirudin and further sperimentations will clarify if the standard safety profile of bivalirudin will be preserved.

DEFINITIONS

Thrombin: a protein that acts as a blood clotting factor

Anticoagulant action: the action to stop the multilevel cascade of blood clotting system.

Direct thrombin inhibitors: inhibitors of thrombin able to act without the intermediation of other factors

Univalent DTIs: binding only one site on thrombin surface

Bivalent DTIs: binding at the active site and at the exosite 1 on thrombin surface.

Acute coronary syndromes: include a spectrum of clinical conditions related to unstable coronary artery disease and ranging from unstable angina to myocardial infarction.

Percutaneous coronary intervention: also known as coronary balloon angioplasty, it is used to revascularize narrowed or occluded coronary arteries through balloons or stents with a percutaneous endoluminal approach.

GPIIb/IIIa Inhibitor: new drugs used to prevent platelet aggregation and thrombus formation in acute coronary syndromes.

Activated clotting time: a test that is used to monitor the effectiveness of anticoagulants.

Deep venous thrombosis: the formation of a clot in the the large veins in the lower leg and thigh.

Major bleedings: they are major life-threatening bleedings (e.g., massive hematuria, gastrointestinal bleeding, retroperitoneal bleeding, or intracranial bleeding), that usually require hemotrasfusion.

MACEs: major adverse cardiac event (death, myocardial infarction, myocardial revascularization) that are usually recorded in the big trials.

Stent thrombosis: clot formation inside the stent after percutaneous coronary intervention.

LIST OF ABBREVIATIONS

ACS	:	acute coronary syndrome
ACT	:	activated clotting time
DTI	:	direct thrombin inhibitors
GPI	:	GP IIb/IIIa Inhibitor
HIT	:	heparin-induced thrombocytopenia
MI	:	myocardial infarction
PCI	:	percutaneous coronary intervention

REFERENCES CITED

A comparison of recombinant hirudin with heparin for the treatment of acute coronary syndromes. 1996. The Global Use of Strategies to Open Occluded Coronary Arteries (GUSTO) IIb investigators. N Engl J Med 335(11): 775–782.

A low molecular weight, selective thrombin inhibitor, inogatran, vs heparin, in unstable coronary artery disease in 1209 patients. 1997. A double-blind, randomized, dose-finding study. Thrombin inhibition in Myocardial Ischaemia (TRIM) study group. Eur Heart J 18(9): 1416–1425.

Bates, S.M. and J.I. Weitz .1998. Direct thrombin inhibitors for treatment of arterial thrombosis: potential differences between bivalirudin and hirudin. Am J Cardiol 82(8B): 12P–18P.

Bittl, J.A., J. Strony, J.A. Brinker, W.H. Ahmed, C.R. Meckel, B.R. Chaitman, J. Maraganore, E. Deutsch and B. Adelman. 1995. Treatment with bivalirudin (Hirulog) as compared with heparin during coronary angioplasty for unstable

or postinfarction angina. Hirulog Angioplasty Study Investigators. N Engl J Med 333(12): 764–769.
Cortese, B., A. Picchi, A. Micheli, A.G. Ebert, F. Parri, S. Severi and U. Limbruno. 2009. Comparison of prolonged bivalirudin infusion versus intraprocedural in preventing myocardial damage after percutaneous coronary intervention in patients with angina pectoris. Am J Cardiol 104(8): 1063–1068.
Coughlin, S.R. 2000. Thrombin signalling and protease-activated receptors. Nature 407(6801): 258–264.
Direct thrombin inhibitors in acute coronary syndromes: principal results of a meta-analysis based on individual patients' data. 2002. Lancet. 359(9303): 294–302.
Doyle, B.J., C.S. Rihal, D.A. Gastineau and D.R. Holmes Jr. 2009. Bleeding, blood transfusion, and increased mortality after percutaneous coronary intervention: implications for contemporary practice. J Am Coll Cardiol 53(22): 2019–2027.
Effects of recombinant hirudin (lepirudin) compared with heparin on death, myocardial infarction, refractory angina, and revascularisation procedures in patients with acute myocardial ischaemia without ST elevation: a randomised trial.1999. Organisation to Assess Strategies for Ischemic Syndromes (OASIS-2) Investigators. Lancet 353(9151): 429–438.
Eriksson, B.I., O.E. Dahl, M.H. Huo, A.A. Kurth, S. Hantel, K. Hermansson, J.M. Schnee and R.J. Friedman. 2011. Oral dabigatran versus enoxaparin for thromboprophylaxis after primary total hip arthroplasty (RE-NOVATE II). A randomised, double-blind, non-inferiority trial. Thromb Haemost 105(4).
Gold, H.K., F.W. Torres, H.D. Garabedian, W. Werner, I.K. Jang, A. Khan, J.N. Hagstrom, T. Yasuda, R.C. Leinbach, J.B. Newell, et al. 1993. Evidence for a rebound coagulation phenomenon after cessation of a 4-hour infusion of a specific thrombin inhibitor in patients with unstable angina pectoris. J Am Coll Cardiol 21(5): 1039–1047.
Herrman, J.P., R. Simon, V.A. Umans, P.F. Peerboom, D. Keane, J.J. Rijnierse, D. Bach, P. Kobi, R. Kerry, P. Close, et al. 1995. Evaluation of recombinant hirudin (CGP 39,393/TMREVASC) in the prevention of restenosis after percutaneous transluminal coronary angioplasty. Rationale and design of the HELVETICA trial, a multicentre randomized double blind heparin controlled study. Eur Heart J 16 Suppl L: 56–62.
Hirsh, J. 2001. New anticoagulants. Am Heart J 142(2 Suppl): S3–8.
Hirsh, J., M. O'Donnell and J.W. Eikelboom. 2007. Beyond unfractionated heparin and warfarin: current and future advances. Circulation 116(5): 552–560.
Kam, P.C., N. Kaur and C.L. Thong. 2005. Direct thrombin inhibitors: pharmacology and clinical relevance. Anaesthesia 60(6): 565–574.
Klootwijk, P., T. Lenderink, S. Meij, H. Boersma, R. Melkert, V.A. Umans, J. Stibbe, E.J. Muller, K.J. Poortermans, J.W. Deckers and M.L. Simoons. 1999. Anticoagulant properties, clinical efficacy and safety of efegatran, a direct thrombin inhibitor, in patients with unstable angina. Eur Heart J 20(15): 1101–1111.
Lane, D.A., H. Philippou and J.A. Huntington. 2005. Directing thrombin. Blood 106(8): 2605–2612.
Lewis, B.E., D.E. Wallis, S.D. Berkowitz, W.H. Matthai, J. Fareed, J.M. Walenga, J. Bartholomew, R. Sham, R.G. Lerner, Z.R. Zeigler, P.K. Rustagi, I.K. Jang, S.D. Rifkin, J. Moran, M.J. Hursting and J.G. Kelton. 2001. Argatroban anticoagulant

therapy in patients with heparin-induced thrombocytopenia. Circulation 103(14): 1838–1843.

Lincoff, A.M., J.A. Bittl, R.A. Harrington, F. Feit, N.S. Kleiman, J.D. Jackman, I.J. Sarembock, D.J. Cohen, D. Spriggs, R. Ebrahimi, G. Keren, J. Carr, E.A. Cohen, A. Betriu, W. Desmet, D.J. Kereiakes, W. Rutsch, R.G. Wilcox, P.J. de Feyter, A. Vahanian and E.J. Topol. 2003. Bivalirudin and provisional glycoprotein IIb/IIIa blockade compared with heparin and planned glycoprotein IIb/IIIa blockade during percutaneous coronary intervention: REPLACE-2 randomized trial. JAMA 289(7): 853–863.

Mehran, R., A.J. Lansky, B. Witzenbichler, G. Guagliumi, J.Z. Peruga, B.R. Brodie, D. Dudek, R. Kornowski, F. Hartmann, B.J. Gersh, S.J. Pocock, S.C. Wong, E. Nikolsky, L. Gambone, L. Vandertie, H. Parise, G.D. Dangas and G.W. Stone. 2009. Bivalirudin in patients undergoing primary angioplasty for acute myocardial infarction (HORIZONS-AMI): 1-year results of a randomised controlled trial. Lancet 374(9696): 1149–1159.

Meyer, B.J., A. Fernandez-Ortiz, A. Mailhac, E. Falk, L. Badimon, A.D. Michael, J.H. Chesebro, V. Fuster and J.J. Badimon. 1994. Local delivery of r-hirudin by a double-balloon perfusion catheter prevents mural thrombosis and minimizes platelet deposition after angioplasty. Circulation 90(5): 2474–2480.

Meyer, B.J., J.J. Badimon, J.H. Chesebro, J.T. Fallon, V. Fuster and L. Badimon. 1998. Dissolution of mural thrombus by specific thrombin inhibition with r-hirudin: comparison with heparin and aspirin. Circulation 97(7): 681–685.

Shantsila, E., G.Y. Lip and B.H. Chong. 2009. Heparin-induced thrombocytopenia. A contemporary clinical approach to diagnosis and management. Chest 135(6): 1651–1664.

Stone, G.W., B.T. McLaurin, D.A. Cox, M.E. Bertrand, A.M. Lincoff, J.W. Moses, H.D. White, S.J. Pocock, J.H. Ware, F. Feit, A. Colombo, P.E. Aylward, A.R. Cequier, H. Darius, W. Desmet, R. Ebrahimi, M. Hamon, L.H. Rasmussen, H.J. Rupprecht, J. Hoekstra, R. Mehran and E.M. Ohman. 2006. Bivalirudin for patients with acute coronary syndromes. N Engl J Med 355(21): 2203–2216.

Topol, E. 2001. Recent advances in anticoagulant therapy for acute coronary syndromes. Am Heart J 142(2 Suppl): S22–29.

Wallentin, L., R.G. Wilcox, W.D. Weaver, H. Emanuelsson, A. Goodvin, P. Nystrom and A. Bylock. 2003. Oral ximelagatran for secondary prophylaxis after myocardial infarction: the ESTEEM randomised controlled trial. Lancet 362(9386): 789–797.

Wallentin, L., S. Yusuf, M.D. Ezekowitz, M. Alings, M. Flather, M.G. Franzosi, P. Pais, A. Dans, J. Eikelboom, J. Oldgren, J. Pogue, P.A. Reilly, S. Yang and S.J. Connolly. 2010. Efficacy and safety of dabigatran compared with warfarin at different levels of international normalised ratio control for stroke prevention in atrial fibrillation: an analysis of the RE-LY trial. Lancet 376(9745): 975–83.

Weitz, J.I., M. Hudoba, D. Massel, J. Maraganore and J. Hirsh. 1990. Clot-bound thrombin is protected from inhibition by heparin-antithrombin III but is susceptible to inactivation by antithrombin III-independent inhibitors. J Clin Invest 86(2): 385–391.

9

Anti-Anginal Drugs in Focus: Trimetazidine

Mario Marzilli[1,]* and *Alda Huqi*[1,2]

ABSTRACT

Not all patients with chronic angina are candidates for coronary revascularization and, even when such a strategy is adopted, not all of them succeed to be symptom free. Importantly, this subset represents an increasing population, particularly among diabetic and the elderly population. Irrespective of the underlying cause, the final common pathophysiological mechanism is a reduced oxygen supply to the myocardium for which revascularization and traditional medical therapies are not sufficient options for symptom control.

One of the mechanisms myocardial ischemia has been shown to reduce cardiac efficiency and give rise to the ischemic cascade is an increase in fatty acid oxidation which results in a relative decrease in glucose oxidation and therefore need for more oxygen for ATP molecules produced and intracellular accumulation of protons. In this scenario, therapeutic interventions aimed at a shift of myocardial substrate utilization towards glucose metabolism would particularly benefit cardiac efficiency and ischemic heart disease (IHD) symptoms. This can be achieved by either inhibiting fatty acid oxidation or stimulating

---
[1]Cardio-Thoracic and Vascular Department, Via Paradisa, 2, 56100–Pisa, Italy; Email: mario.marzilli@med.unipi.it
[2]Mazankowski Alberta Heart Institute, T6G 2S2, Edmonton, Alberta, Canada; Email: alda_h@hotmail.com; huqi@ualberta.ca
*Corresponding author

List of abbreviations after the text.

glucose oxidation. Trimetazidine, a partial inhibitor of fatty acid oxidation, induces a shift from free fatty acid towards predominantly glucose utilization and has been shown to confer substantial symptom relief in different treatment regimens of patients with stable chronic angina. In fact the drug may be added to the ongoing therapy with β-blockers (BBs), calcium channel blockers (CCBs), and nitrates with safety and, has been shown to have a high safety and tolerability profile, in the absence of known drug interactions. However, although the safety and clinical benefits have been documented since the early 80s, it still lacks a widespread clinical use or a guideline recommendation in the management of chronic stable angina patients.

INTRODUCTION

Coronary revascularization procedures by means of percutaneous coronary interventions (PCI) or coronary artery bypass graft surgeries (CABG) are performed daily worldwide for the treatment of patients with myocardial ischemia. Nevertheless, angina remains a significant clinical problem. In fact, large clinical trials consistently report that many chronic angina patients present with persistent symptoms after coronary revascularization (Table 9.1) (Boden et al. 2007). Of those, as many as two thirds might require one or more classical anti-angina agents (Holubkov et al. 2002) and, yet

Table 9.1 Percentage of patients with symptoms persistence/reoccurrence in some of the large clinical trials.

Study	Follow up duration	Type of intervention	% patients with angina in the different treatment groups
Mass-II	1 yr	CABG, PCI, MT	12% CABG, 21% PCI, 54% OMT
RITA-2	1 yr	PCI, MT	38% PCI, 57% MT
BARI	1 yr	CABG, PCI	10% CABG, 30% PCI
COURAGE	5 yr	PCI, MT	26% PCI, 28% MT
FAME	2 yr	Angiography-guided PCI, FFR-guided PCI	24% angiography- guided PCI, 20% FFR-guided PCI

PCI (percutaneous coronary intervention), CABG (coronary artery bypass grafting), FFR (fractional flow reserve), MT (medical therapy).

the results are not optimal. Revascularization procedures (either percutaneous or surgical) are performed with the aim of removing the flow limiting stenosis from the epicardial coronary artery. This rationale is in line with the main pathophysiological mechanism underlying angina (epicardial stenosis) and the benefit obtained in the majority of patients undergoing revascularization is a witness. However, there are three main groups of angina patients to be considered: 1. Those who are deemed to be unsuitable for coronary revascularization because of diffuse coronary artery disease (CAD) (refractory angina). Importantly, this subset represents an increasing population, particularly among diabetic and elderly patients; 2. Those who experience recurrent angina after percutaneous or surgical coronary procedures (recurrent angina) for which, several potential causes, such as bypass graft failure, restenosis, or atherosclerotic disease progression have been identified as the cause of symptom reoccurrence; 3. Those who undergo successful revascularization, in which none of the above mentioned factors can be identified as the cause of symptom reoccurrence (persistent angina). In this latter group, factors other than epicardial stenosis, such as microvascular dysfunction have been suggested as the underlying physiopathological mechanism for persistent symptoms.

While the first condition is prevalently associated with poor quality of life, the latter two are also associated with repeat invasive procedures, increased patient/physician frustration and, last but not least, increased healthcare costs. However, irrespective of the underlying cause, the final common pathophysiological mechanism is a reduced oxygen supply to the myocardium for which, both revascularization and traditional medical therapies are not sufficient for controlling symptoms. Considerable progress has been made over the last 30 yr in expanding the therapeutic options for ischemic heart disease (IHD). Among these, metabolic modulation therapy by means of drug agents such as trimetazidine has been shown to confer significant symptom relief. The drug may be added to ongoing therapy with β-blockers (BBs), calcium channel blockers (CCBs), and nitrates with safety and, has been shown to have a high safety and tolerability profile, in the absence of known drug interactions (Chazov et al. 2005).

CARDIAC METABOLISM AND RATIONAL FOR METABOLIC MODULATION THERAPY

Glycolysis and mitochondrial oxidative metabolism are the principal sources of energy production in the adult heart, with the latter accounting for more than 95% of the total amount. In normal condition, 50–70% of the energy produced by mitochondrial oxidative metabolism is recovered from fatty acid β-oxidation. The remaining 30–50% is recovered from the oxidative metabolism of the glycolytic products of glucose (pyruvate).

Relevance of Metabolism Alterations in IHD

In aerobic conditions, despite producing more adenosine triphosphate (ATP) than carbohydrates (in terms of ATP per gram of substrate), fatty acids require about 10–15% more oxygen to produce an equivalent amount of ATP (Stanley et al. 2005). Therefore, in terms of oxygen consumption, oxidation of a molecule of glucose represents a more efficient energy production pathway. Such a property becomes particularly relevant in conditions of reduced oxygen supply such as IHD.

As mentioned, glucose metabolism begins with glycolysis, a cytosolic process that converts glucose to pyruvate in the cytosol. In this way, glycolysis provides pyruvate to the pyruvate dehydrogenase enzyme (PDH) which is the rate limiting enzyme for glucose oxidation (Stanley et al. 2005). When PDH is active, pyruvate enters the mitochondria and is converted to acetyl-CoA, therefore fueling the Krebs cycle for ATP production. Conversely, fatty acid oxidation, which occurs within the mitochondria, is another source of acetyl-CoA for the Krebs cycle (Figs. 9.1 and 9.2). These two sources of acetyl-CoA are highly dependent on and affected from each other, a phenomenon known as the 'Randle cycle' (Randle et al. 1963). For example, a high level of fatty acids which stimulate an increase in fatty acid oxidation, can indirectly inhibit the oxidation of carbohydrates through inhibition of PDH activity (Depre et al. 1999). As a consequence pyruvate cannot enter the mitochondria and therefore cannot fuel oxidative metabolism. This results in the production of only 2 ATP per glucose molecule (versus 36 ATP during glucose oxidation) and conversion of pyruvate to lactate with proton accumulation.

Anti-Anginal Drugs in Focus: Trimetazidine   159

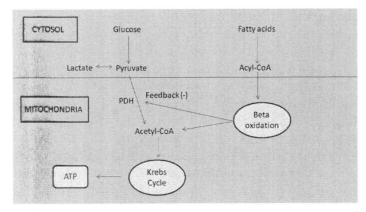

Figure 9.1 Overview of glucose and fatty acid metabolism.

Glucose is converted into pyruvate in the cytosol, converted into acetyl-CoA by PDH in the mitochondria and then shuttled into the Krebs cycle for ATP production. Fatty acids are activated in fatty acyl-CoA and then transferred into the mitochondria where they undergo beta oxidation for production of acetyl-CoA. Acetyl-CoA produced from mitochondria enters Krebs cycle but can also inhibit PDH and therefore glucose metabolism. Abbreviations: PDH (pyruvate dehydrogenase), ATP (adenosine triphosphate), LC-3KAT (long-chain 3-ketoacyl CoA thiolase).

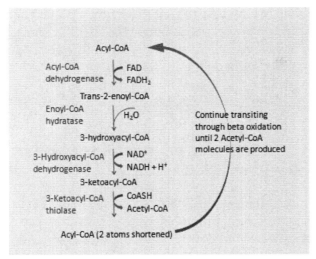

Figure 9.2 Fatty acid beta-oxidation.

Each cycle of the Acyl-CoA (activated fatty acid) is cleaved two carbons in length, therefore producing (n-2) Acyl-CoA + Acetyl-CoA. The cycle is repeated until the final products are two molecules of Acetyl-CoA. Left (blue) the four enzymes involved in beta oxidation process. Abbreviations: FAD (flavine adenine dinucleotide), FADH2 (reduced FAD), NAD (nicotinamide adenine dinucleotide), NADH (reduced NAD).

A primarily effect of ischemia is an overall reduced ATP formation in mitochondria. This triggers accelerated glycolysis and reduced cell pH, calcium accumulation, potassium efflux and adenosine formation (Liu et al. 2002). Moreover, pain resulting from myocardial ischemia leads to enhanced catecholamine release and increased lipolysis. This condition is associated with an increase in circulating fatty acids levels, a relative increase in fatty acid oxidation and therefore (through the 'Randle cycle') a reduced glucose oxidation rate. Myocardial ischemia has therefore a prominent role in increasing the uncoupling between glycolysis and glucose oxidation by contemporarily increasing glycolysis and reducing glucose oxidation rates, therefore further reducing cardiac efficiency. The need to use ATP for reestablishing ionic homeostasis instead of supporting contractile function is another cause for reduced cardiac efficiency. In addition, intracellular proton accumulation also directly decreases the efficiency of the contractile proteins and, as such, cardiac efficiency (El Banani et al. 2000).

Therefore, therapeutic interventions aimed at shifting myocardial substrate utilization from fatty acid towards glucose metabolism would particularly benefit cardiac efficiency and IHD symptoms. Given the interdependence between fatty acid and glucose oxidation this can be achieved by either inhibiting fatty acid oxidation or stimulating glucose oxidation. In this chapter we will focus on the use of trimetazidine, a partial inhibitor of fatty acid oxidation, in patients with stable chronic angina.

Beneficial Cellular Effects of Fatty Acid Oxidation Inhibition by Trimetazidine

There are numerous ways to inhibit cardiac fatty acid oxidation, some of which include the inhibition of fatty acid transport into the cardiac myocyte, the inhibition of fatty acid uptake into the mitochondria, and the inhibition of the enzymatic machinery of the β-oxidative pathway itself.

Trimetazidine (1-[2,3,4-trimethoxybenzyl] piperazine dihydrochloride) (Fig. 9.3) is metabolic modulator agent that has been used for more than two decades in Europe. It inhibits fatty acid oxidation by blocking the β-oxidative enzyme, long-chain

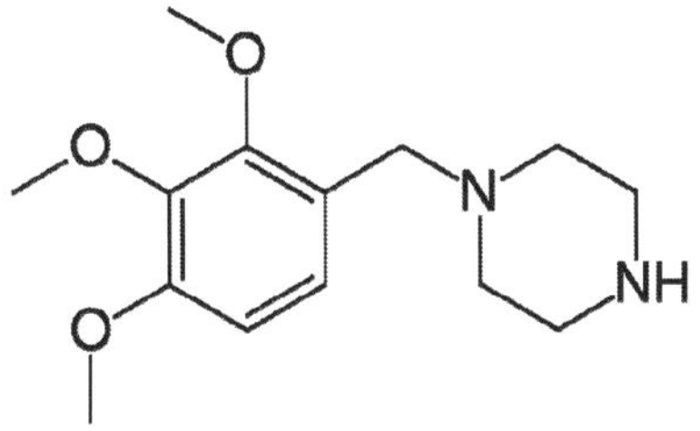

Figure 9.3 Chemical structure of trimetazidine.

3-ketoacyl CoA thiolase (LC 3-KAT) (Kantor et al. 2000). This effect results in an increase in PDH activity which compensates the reduced availability of acetyl-CoA derived from β-oxidation of fatty acids, by providing them through glucose oxidation. In this way, stimulation of glucose oxidation increases glycolysis/glucose oxidation coupling, resulting in a decreased proton production, a decrease in tissue acidosis, intracellular calcium overload and free radical production. Finally, these chain effects prevent a significant decrease in ATP and phosphocreatine levels in response to hypoxia or ischemia, preserve ionic pump function and therefore translate into an improved cardiac efficiency (Kantor et al. 2000) and reduced symptoms. Importantly, these effects are not associated with significant alterations in hemodynamic parameters, which are particularly important to patients who are already on anti-ischemic treatment with hemodynamic agents.

CLINICAL EFFICACY OF TRIMETAZIDINE

Treatment with trimetazidine has been shown to confer beneficial effects in various clinical forms of ischemia including angina pectoris, acute coronary syndromes (ACS), as well as in heart failure trials.

In patients with chronic angina, trimetazidine increases work capacity and delays the appearance of symptoms and ECG changes during exercise (Sellier et al. 1987, Detry 1993, Sellier 1986). Importantly, the benefits observed after acute administration are maintained in chronic treatment, which is well tolerated by patients (Passeron 1986). In fact, it exhibits no significant negative inotropic or vasodilator properties either at rest or during dynamic exercise (Pornin et al. 1994).

The efficacy of trimetazidine as an anti-anginal drug has been assessed in randomized, placebo-controlled studies, both as 'solo' treatment and in combination or comparison with BBs and CCBs (Table 9.2).

Trimetazidine as Monotherapy

The beneficial effects of trimetazidine as an anti-anginal drug were first tested as monotherapeutic regimens in chronic angina patients who were withdrawn from all anti-angina medications at least 8 d before inclusion (Passeron 1986, Sellier 1986). The results showed a significant improvement in exercise capacity (increased total work and increase in the duration of exercise) in response to trimetazidine treatment. Similarly, there was a significant reduction in the weekly number of angina attacks in the trimetazidine group as compared with placebo. No differences in blood pressure (BP) and heart rate (HR) were observed between the two groups, confirming the absence of any hemodynamic effect of trimetazidine.

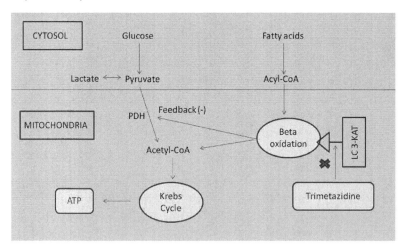

Figure 9.4 Mechanism of action of trimetazidine.
Trimetazidine inhibits LC-3KAT, one of the enzymes of fatty acid oxidation. In this way there is a reduced production of acetyl-CoA originating from fatty acid oxidation and therefore a relief of PDH activity. Abbreviations: PDH (pyruvate dehydrogenase), ATP (adenosine triphosphate), LC-3KAT (long-chain 3-ketoacyl CoA thiolase) Kantor 2000.

Anti-Anginal Drugs in Focus: Trimetazidine 163

Table 9.2 Principal trimetazidine studies considered in this chapter.

Author	Year	Study title
	TMZ as monotherapy	
Sellier et al.	1986	The effects of TMZ on ergometric parameters of exercise-induced angina.
Passeron et al.	1994	Effectiveness of TMZ in stable effort angina due to chronic coronary insufficiency.
TMZ versus CCBs		
	TMZ with BBs	
		Combination treatment in stable effort angina using TMZ and netoprolol. TRIMPOL II study.
Szwed et al.	2001	Combination of TMZ with nifedipine in effort angina.
	TMZ with CCBs	
Monpere et al.	1990	Combination treatment with TMZ and diltiazem in stable angina pectoris.
Manchanda et al.	1997	Comparison between TMZ and mononitrate isosorbide for patients receiving BBs.
Hanania et al.	*TMZ versus nitrates* 2002	TMZ: a new concept in the treatment of angina. Comparison with propranolol in patients with stable angina.
Detry et al.	*TMZ versus BBs* 1994	Comparison of TMZ with nifedipine in effort angina: a double-blind, crossover study.
Dalla-Volta et al.	*TMZ versus CCBs* 1990	TMZ in geriatric patients with stable angina pectoris: the tiger study.
Kolbel et al.	*TMZ in elderly and diabetic patients*	
Marazzi et al.	2003	Effect of TMZ on quality of life in elderly patients with ischemic dilated cardiomyopathy.
Ribiero et al.	2009	TMZ added to combined hemodynamic antianginal therapy in patients type 2 diabetes: a randomized crossover trial.
Iyengar et al.	2007	Effect of antianginal drugs in stable angina on predicted mortality risk after surviving a myocardial infarction: a preliminary study (METRO).
	Prognostic effects of TMZ 2009	

TMZ (trimetazidine), BBs (β-blockers), CCBs (calcium channel blockers).

Trimetazidine in Combination with BBs

TRIMPOL II was a randomized, double-blind, placebo-controlled, multicenter study that recruited 426 patients with stable angina who were randomized to either trimetazidine 20 mg three times a day or placebo on top of metoprolol (Szwed et al. 2001). The addition of trimetazidine to standard therapy with metoprolol resulted in improvement in the time to ST segment depression on exercise tolerance testing, total exercise workload, mean nitrate consumption, and angina frequency as compared to patients receiving placebo. Moreover the drug had a favorable side-effect profile with the most common adverse events being nausea, vomiting, fatigue, and myalgia.

Trimetazidine in Combination with CCBs

A first study evaluated the addition of trimetazidine as compared to placebo in patients with a persistently positive stress test, despite at least 15 d of treatment with nifedipine (Monpere et al. 1990). Trimetazidine use was associated with an increase in maximal workload, while this parameter remained stable or even slightly deteriorated with placebo. Moreover, mean weekly frequency of angina attacks decreased with the use of trimetazidine as compared to placebo.

Subsequent studies of combination therapy with non-dihydropyridines confirmed the beneficial effects of a combination treatment with trimetazidine, in the absence adverse hemodynamic events or increased side effects (Manchanda and Krishnaswami 1997).

Trimetazidine versus Nitrates

Hanania et al. (2002) investigated the efficacy of trimetazidine as compared to isosorbide mononitrate in symptomatic angina patients, despite a daily regimen of 100 mg of atenolol. Both drugs induced significant and comparable clinical benefits in terms of improvement in quality of life and reduced positive exercise stress test results. However, as opposed to nitrate use, the addition of trimetazidine to treatment regimen was not associated with significant hemodynamic or other adverse effects such as chepalalgia.

Trimetazidine versus BBs

The effects of trimetazidine (20 mg three times daily) were compared with those of propranolol (40 mg three times daily) in a double-blind parallel group multicenter study in 149 men with stable angina (Detry et al. 1994). After 3 mon of treatment, similar anti-angina efficacy was observed between the trimetazidine and propranolol groups. No significant differences with regard to angina frequency, exercise duration or time to 1 mm ST segment depression were observed between the different treatment groups. However, while HR and rate x pressure product at rest and at peak exercise decreased with propranolol, no significant changes were observed in the trimetazidine group.

Trimetazidine versus CCBs

Trimetazidine efficacy has also been shown in a head to head, double-blind study with the dihydropyridynic agent nifedipine (Dalla-Volta et al. 1990). The results show that nifedipine and trimetazidine both decreased the number of angina attacks and increased workload parameters without any significant difference between the two drugs. However, at a constant level of work, the rate x pressure product decreased with nifedipine, but remained stable with trimetazidine.

Trimetazidine in Elderly and Diabetic Patients

As mentioned, the aging population and diabetic patients frequently present with refractory angina because of diffuse atherosclerotic disease or other co-morbidities. The use of trimetazidine has been shown to benefit both these patient populations. In the TIGER study (Kolbel and Bada 2003) involving 141 stable angina patients aged 65–86 years, trimetazidine was shown to improve exercise stress tests and angina symptoms. Because of its metabolic effects, free from any hemodynamic action, trimetazidine proved to be beneficial in elderly patients, with an excellent tolerance profile. Another study assessed the effects of trimetazidine in addition to standard cardiovascular therapy on left ventricular function and quality of life (QOL) parameters in elderly patients with ischemic heart disease and reduced left ventricular function (Marazzi et al. 2009).This was a randomized placebo controlled study involving 47 elderly symptomatic patients who were already on optimal medical therapy (OMT). At six mon after randomization, patients on trimetazidine showed a significant improvement in clinical conditions and QOL.

A significant improvement in QOL and exercise capacity has been also observed in diabetic chronic angina patients presenting with coronary anatomy not amenable to revascularization who were treated with trimetazidine on top of OMT (Ribeiro et al. 2007).

Potential Prognostic Impact of Trimetazidine on Mortality of Angina Patients

The METRO (ManagEment of angina: a reTRospective cOhort) study assessed the effect that different anti-angina drugs had on subsequent (6 mon) mortality risk of patients with stable angina experiencing a myocardial infarction (Iyengar and Rosano 2009). Amongst the use of at least one anti-anginal drug (nitrates, beta-adrenoceptor antagonists, calcium channel antagonists, trimetazidine, or nicorandil) over several months prior to a myocardial infarction (MI), use of trimetazidine was the only agent to be associated with a mortality benefit.

In accordance with the previously discussed study the, Cochrane review (Ciapponi et al. 2005), which included randomized studies comparing trimetazidine with placebo, or other anti-anginal drugs in adults with stable angina in a total of 1,378 patients, found that the drug is effective in the treatment of stable angina compared with placebo, alone or combined with conventional anti-anginal agents. Moreover, trimetazidine use resulted in fewer dropouts due to adverse events.

However, although the clinical benefits have been documented since the early 80s, unfortunately trimetazidine still lacks a widespread clinical use or a guideline recommendation in the management of chronic stable angina patients.

PRACTICE AND PROCEDURES

Pharmacokinetic Properties of Immediate- and Modified-release Trimetazidine

Trimetazidine dihydrochloride is freely soluble in water and has two pKa values 4.32 and 8.95. Immediate release preparation is administered orally in divided doses of 40 to 60mg daily. It is quickly absorbed and eliminated by the organism with plasma half-life of around 6.0 +/−1.4 hr and Tmax of around 1.8 +/−0.7 hr. Since it has a shorter plasma half life, in practice 20 mg preparation is given twice or thrice a day in order to ensure relatively constant plasma levels. This was also the reason for the manufacture of the modified-release dosage form containing 35 mg for twice- daily administration. This dosage form is bioequivalent to the 20 mg conventional thrice-a-day formulation of trimetazidine hydrochloride. The modified-release formulation is based on a hydrophilic matrix that utilizes polymers which swell in contact with gastrointestinal fluids to form gels that are subsequently gradually absorbed.

KEY FACTS

Trimetazidine in other Clinical Scenarios

- The benefits of trimetazidine use are not limited to only patients with chronic stable angina.
- In fact its use has been shown to be beneficial in other settings such as ACS, heart failure, hypertrophic cardiomyopathy (HCM).
- Trimetazidine has been shown to confer benefits in terms of reduced ischemia-reperfusion injury when administered intravenously in the ACS setting. However these findings have not always been reproduced in clinical testing and have been mainly restricted to patients not undergoing revascularization.
- Emerging evidence suggests that the failing heart preferentially utilizes fatty acid oxidation at the expense of glucose oxidation.

- Such findings are supported by the clinical benefits obtained with trimetazidine in terms of both improved ejection fraction and functional class in patients with heart failure.
- There is also increasing evidence that impairment of myocardial energetics occurs in patients with HCM, so that energetic impairment has been proposed to play a primary role in the pathophysiology of HCM.
- Metabolic modulation therapy has been in fact shown to be beneficial also in this patient group.

Perhexiline: a 'historical' Fatty Acid Oxidation Inhibitor

- Perhexiline is another metabolic agent that inhibits fatty acid oxidation through inhibition of carnitine palmitoyltransferase-1 (CPT-1), a key enzyme in the transport of fatty acids into the mitochondria for oxidation.
- Given the interdependence between fatty acid and glucose metabolism, this effect translates in increased glucose oxidation rates, increased cardiac efficiency and therefore amelioration of ischemic symptoms.
- However, in spite of the proven benefits in patients with IHD, its use has been limited because of adverse effects such as neurotoxicity and hepatotoxicity during long-term therapy in slow drug metabolizers.
- Nonetheless dose titration and monitoring prevent the hepatotoxicity and neuropathy, in this way allowing its clinical benefits in various clinical settings such as refractory angina, aortic stenosis and heart failure.
- Although the pharmacological effects of perhexiline are due to an alternative mechanism (inhibition of CPT-1) its clinical efficacy further point out the role of fatty acid inhibition (and therefore of glucose oxidation stimulation) in augmentation of cardiac efficiency and therefore amelioration of IHD symptoms.

Malonyl-CoA Decarboxylase (MCD) Inhibitors: Future Metabolic Agents

- Malonyl-CoA is a potent endogenous inhibitor of CPT-1, the rate-limiting enzyme in the mitochondrial uptake of fatty acids.

- Thus, malonyl-CoA decreases the uptake of fatty acids into the mitochondria, thereby reducing mitochondrial fatty acid β-oxidation.
- Malonyl-CoA decarboxylase (MCD) degrades malonyl-CoA and this effects leads to an increased fatty acid oxidation.
- Inhibition of MCD significantly increases malonyl-CoA levels, therefore causing a significant decrease in fatty acid oxidation rates and a subsequent increase in glucose oxidation rates.
- In line with such pharmacodynamic properties, animal models of inhibition of MCD have shown a significant improvement in cardiac functional recovery of aerobically reperfused ischaemic hearts.
- Inhibition of MCD in the heart appears to be a safe and a very promising therapeutic target for IHD.

SUMMARY POINTS

- Persistent/refractory angina and recurrent chest pain after coronary revascularization are always a disappointment to both the patient and the cardiologist.
- Given the continuously increasing aging population with associated co-morbidities, this patient population is expected to become more and more represented.
- Currently available (and used) treatment strategies (revascularization and hemodynamic drug agents) may not always offer a valid solution for symptom relief in this patient population.
- 'Metabolic' anti-angina drugs induce a shift from free fatty acid towards, predominantly, glucose utilization by the myocardium to increase ATP generation per unit of oxygen consumed.
- Metabolic agents therefore improve effectiveness of energy production, decrease the oxygen debt, and protect myocardial cells from the effects of ischemia.
- In this way they provide a valid alternative action mechanism as compared to classical hemodynamic agents which induce changes such as reduction in systemic vascular resistance, coronary vasodilatation, or negative inotropism.

- These effects offer particular advantage in patients in whom conventional agents may induce symptomatic hypotension, inappropriate bradycardia, or worsening heart failure.
- Trimetazidine, a fatty acid oxidation inhibitor, is the most clinically tested metabolic drug, which has been shown to confer benefits both when used as only an anti-angina agent, and on top of other classical agents.
- The drug may be added to ongoing therapy with BBs, CCBs, and nitrates with safety and, has been shown to have a high safety and tolerability profile, in the absence of known drug interactions.
- However, although the clinical benefits have been documented since the early 80s, unfortunately it still lacks widespread clinical use or a guideline recommendation in the management of chronic stable angina patients.

DEFINITIONS

Fatty acid oxidation: this process is termed β-oxidation since it occurs through the sequential removal of 2-carbon units by oxidation at the β-carbon position of the fatty acyl-CoA molecule. Each round of β-oxidation produces one mole of NADH, one mole of $FADH_2$ and one mole of acetyl-CoA which then enters the Krebs cycle.

Glycolysis: is the biochemical pathway that initiates the oxidation of glucose. Glycolysis occurs in the cytoplasm and it splits the six-carbon glucose molecule into two three-carbon molecules of pyruvate. This is accomplished through a series of chemical reactions during which a small amount of ATP and two molecules of pyruvate are formed.

Glucose oxidation: pyruvate formed in glycolysis is transported into the mitochondria converted to acetyl-CoA by PHD enzyme. Consequently acetyl-CoA originating from glucose exhibits the same fate of that formed from fatty acid beta oxidation: it enters the Krebs cycle for energy production.

Krebs cycle: is a series of enzyme-catalysed chemical reactions that occurs in the matrix of the mitochondrion. Krebs cycle is part of a fundamental metabolic pathway involved in the chemical conversion of carbohydrates, fats and proteins (namely their acetyl-CoA products) into carbon dioxide and water to generate a form of usable energy such as ATP.

Pyruvate: is the three carbon end product of glycolysis (two puruvates for each glucose molecule), which is converted into acetyl-CoA. When the oxygen is insufficient, pyruvate is broken down anaerobically, creating lactate.

Acetyl-CoA: is the activated acetate, which is composed of two carbon atoms. This important coenzyme is the metabolic product of the oxidation of several amino acids, pyruvate and fatty acids. The acetyl-CoA is then broken down and used by the Krebs cycle for energy production.

Cardiac efficiency: As for a machine, the efficiency of the heart is the ratio of effective work to the energy expended in producing it, i.e., energy spent for contractile purposes/energy spent for contractile purposes + energy spent for metabolic purposes + energy spent for ionic homeostasis + energy spent for heat production.

Persistent angina: lack of angina symptom relief for more than 6 mon after a coronary revascularization procedure.

Refractory angina: angina pectoris not be amenable any more to standard treatment (revascularization and hemodynamic drug therapy).

Microvascular dysfunction: presence of signs and/or symptoms of myocardial ischemia in the absence for which microvascular abnormalities (increased distal resistances, as assessed by invasive coronary angiography) rather than epicardial coronary stenosis can be demonstrated.

LIST OF ABBREVIATIONS

ACS	:	acute coronary syndromes
ATP	:	adenosine triphosphate
BBs	:	beta blockers
BP	:	blood pressure
CABG	:	coronary artery bypass grafting
CAD	:	coronary artery disease
CCBs	:	calcium channel blockers
CPT-1	:	carnitine palmitoyltransferase-1
HCM	:	hypertrophic cardiomyopathy
HR	:	heart rate
IHD	:	ischemic heart disease

LC 3-KAT   :   long-chain 3-ketoacyl CoA thiolase
MCD   :   Malonyl-CoA Decarboxylase
MI   :   myocardial infarction
OMT   :   optimal medical therapy
PCI   :   percutaneous coronary intervention
QOL   :   quality of life

REFERENCES CITED

Boden, W.E., R.A. O'Rourke, K.K. Teo, P.M. Hartigan, D.J. Maron, W.J. Kostuk, M. Knudtson, M. Dada, P. Casperson, C.L. Harris, B.R. Chaitman, L. Shaw, G. Gosselin, S. Nawaz, L.M. Title, G. Gau, A.S. Blaustein, D.C. Booth, E.R. Bates, J.A. Spertus, D.S. Berman, G.B. Mancini and W.S. Weintraub. 2007. Optimal medical therapy with or without PCI for stable coronary disease. N Engl J Med 356(15): 1503–1516.

Chazov, E.I., V.K. Lepakchin, E.A. Zharova, S.B. Fitilev, A.M. Levin, E.G. Rumiantzeva and T.B. Fitileva. 2005. Trimetazidine in Angina Combination Therapy—the TACT study: trimetazidine versus conventional treatment in patients with stable angina pectoris in a randomized, placebo-controlled, multicenter study. Am J Ther 12(1): 35–42.

Ciapponi, A., R. Pizarro and J. Harrison. 2005. Trimetazidine for stable angina. Cochrane Database Syst Rev (4): CD003614.

Dalla-Volta, S., G. Maraglino, P. Della-Valentina, P. Viena and A. Desideri. 1990. Comparison of trimetazidine with nifedipine in effort angina: a double-blind, crossover study. Cardiovasc Drugs Ther 4 Suppl 4: 853–859.

Depre, C., J.L. Vanoverschelde and H. Taegtmeyer. 1999. Glucose for the heart. Circulation 99(4): 578–588.

Detry, J.M. 1993. Clinical features of an anti-anginal drug in angina pectoris. Eur Heart J 14 Suppl G: 18–24.

Detry, J.M., P. Sellier, S. Pennaforte, D. Cokkinos, H. Dargie and P. Mathes. 1994. Trimetazidine: a new concept in the treatment of angina. Comparison with propranolol in patients with stable angina. Trimetazidine European Multicenter Study Group. Br J Clin Pharmacol 37(3): 279–288.

El Banani, H., M. Bernard, D. Baetz, E. Cabanes, P. Cozzone, A. Lucien and D. Feuvray. 2000. Changes in intracellular sodium and pH during ischaemia-reperfusion are attenuated by trimetazidine. Comparison between low- and zero-flow ischaemia. Cardiovasc Res 47(4): 688–696.

Hanania, G., R. Haiat, T. Olive, B. Maalouf, D. Michel, M. Martelet and S. Godard. 2002. Coronary artery disease observed in general hospitals: ETTIC study. Comparison between trimetazidine and mononitrate isosorbide for patients receiving betablockers. Ann Cardiol Angeiol (Paris) 51(5): 268–274.

Holubkov, R., W.K. Laskey, A. Haviland, J.C. Slater, M.G. Bourassa, H.A. Vlachos, H.A. Cohen, D.O. Williams, S.F. Kelsey and K.M. Detre. 2002. Angina 1 year after percutaneous coronary intervention: a report from the NHLBI Dynamic Registry. Am Heart J 144(5): 826–833.

Iyengar, S.S., G.M. Rosano. 2009. Effect of antianginal drugs in stable angina on predicted mortality risk after surviving a myocardial infarction: a preliminary study (METRO). Am J Cardiovasc Drugs 9(5): 293–297.

Kantor, P.F., A. Lucien, R. Kozak and G.D. Lopaschuk. 2000. The antianginal drug trimetazidine shifts cardiac energy metabolism from fatty acid oxidation to glucose oxidation by inhibiting mitochondrial long-chain 3-ketoacyl coenzyme A thiolase. Circ Res 86(5): 580–588.

Kolbel, F. and V. Bada. 2003. Trimetazidine in geriatric patients with stable angina pectoris: the tiger study. Int J Clin Pract 57(10): 867–870.

Liu, Q., J.C. Docherty, J.C. Rendell, A.S. Clanachan and G.D. Lopaschuk. 2002. High levels of fatty acids delay the recovery of intracellular pH and cardiac efficiency in post-ischemic hearts by inhibiting glucose oxidation. J Am Coll Cardiol 39(4): 718–725.

Manchanda, S.C. and S. Krishnaswami. 1997. Combination treatment with trimetazidine and diltiazem in stable angina pectoris. Heart 78(4): 353–357.

Marazzi, G., O. Gebara, C. Vitale, G. Caminiti, M. Wajngarten, M. Volterrani, J.A. Ramires, G. Rosano and M. Fini. 2009. Effect of trimetazidine on quality of life in elderly patients with ischemic dilated cardiomyopathy. Adv Ther 26(4): 455–461.

Monpere, C., M. Brochier, J. Demange, G. Ducloux and J.F. Warin. 1990. Combination of trimetazidine with nifedipine in effort angina. Cardiovasc Drugs Ther 4 Suppl 4: 824–825.

Passeron, J. 1986. Effectiveness of trimetazidine in stable effort angina due to chronic coronary insufficiency. A double-blind versus placebo study. Presse Med 15(35): 1775–1778.

Pornin, M., C. Harpey, J. Allal, P. Sellier and P. Ourbak. 1994. Lack of effects of trimetazidine on systemic hemodynamics in patients with coronary artery disease: a placebo-controlled study. Clin Trials Metaanal 29(1): 49–56.

Randle, P.J., P.B. Garland, C.N. Hales and E.A. Newsholme. 1963. The glucose fatty-acid cycle. Its role in insulin sensitivity and the metabolic disturbances of diabetes mellitus. Lancet 1(7285): 785–789.

Ribeiro, L.W., J.P. Ribeiro, R. Stein, C. Leitao and C.A. Polanczyk. 2007. Trimetazidine added to combined hemodynamic antianginal therapy in patients with type 2 diabetes: a randomized crossover trial. Am Heart J 154(1): 78 e71–77.

Sellier, P. 1986. The effects of trimetazidine on ergometric parameters in exercise-induced angina. Controlled multicenter double blind versus placebo study. Arch Mal Coeur Vaiss 79 (9): 1331–1336.

Sellier, P., P. Audouin, B. Payen, P. Corona, T.C. Duong and P. Ourbak. 1987. Acute effects of trimetazidine evaluated by exercise testing. Eur J Clin Pharmacol 33(2): 205–207.

Stanley, W.C., F.A. Recchia and G.D. Lopaschuk. 2005. Myocardial substrate metabolism in the normal and failing heart. Physiol Rev 85(3): 1093–1129.

Szwed, H., Z. Sadowski, W. Elikowski, A. Koronkiewicz, A. Mamcarz, W. Orszulak, E. Skibinska, K. Szymczak, J. Swiatek and M. Winter. 2001. Combination treatment in stable effort angina using trimetazidine and metoprolol: results of a randomized, double-blind, multicentre study (TRIMPOL II). TRIMetazidine in POLand. Eur Heart J 22(24): 2267–2274.

10

Enhanced External Counterpulsation Therapy in Coronary Artery Disease Management

Ozlem Soran[1],* and Debra L. Braverman[2]

ABSTRACT

Coronary artery disease is a narrowing or blockage of the arteries and vessels that provide oxygen and nutrients to the heart. Despite panoply of recent therapeutic advances, patients with coronary artery disease are not adequately treated; therefore, scientists have been investigating new technologies to help these patients. Although the concept of counterpulsation was introduced in the United States in the early 1950s, it took more than 40 yr for investigators to develop the effective technology that is currently being used known as Enhanced External

---

[1]Associate Professor of Medicine, Associate Professor of Epidemiology/Research, University of Pittsburgh Cardiovascular Institute, 200 Lothrop Street, UPMC, Presbyterian Hospital, PUH, F-748, Pittsburgh, PA, 15213; Email: soranzo@upmc.edu
[2]Clinical Associate Professor of Rehabilitation Medicine, Clinical Associate Professor of Medicine, Thomas Jefferson University, Philadelphia, PA, Director of EECP, Co-Director of Cardiac Rehabilitation, Division of Cardiology, Albert Einstein Medical Center, 700 Cottman Avenue, Building B, Lower Level, Philadelphia, PA 19111; Email: bravermd@einstein.edu
*Corresponding author

List of abbreviations after the text.

Counterpulsation (EECP). EECP is a noninvasive method of assisting the circulation, which enhances diastolic augmentation and systolic unloading by means of a pressurized air cuff around the patient's legs that is maintained at 220–300mmHg during diastole. Recent evidence suggests that EECP therapy may improve symptoms and decrease long-term morbidity via several mechanisms, including improvement in endothelial function, promotion of collateralization, enhancement of ventricular function, improvement in oxygen consumption (VO2), regression of atherosclerosis, and peripheral 'training effects' similar to exercise. Numerous clinical trials in the past two decades have shown EECP therapy to be safe and effective for patients with angina, with a clinical response rate averaging 70 to 80%, which is sustained up to 5 yr. The technique of counterpulsation, studied for almost half a century, is considered a safe, highly beneficial, low-cost, noninvasive treatment for these patients with coronary artery disease.

This chapter summarizes the current evidence to support the role of EECP therapy in coronary artery disease management.

INTRODUCTION

According to the American Heart Association and American Stroke Association's 2010 publication on heart disease and stroke statistics, cardiovascular disease remains the leading cause of mortality in the United States in men and women of every major ethnic group. It accounts for nearly 1 million deaths per year as of 2006 and was responsible for one in almost three deaths in the United States in 2006. Approximately 17 million persons have a history of coronary artery disease and 8 million have suffered a myocardial infarction (Lloyd-Jones et al. 2010). There are three common approaches for treating Coronary Artery Disease (CAD): Medication; Percutaneus coronary interventions (PCI: stent implantation or balloon angioplasty); Coronary Bypass Surgery (CABG). It is important to note that none of these approaches provides a cure. In other words, although the symptoms are eliminated or alleviated, the disease and its causes are still present after treatment and require that the patient modify his or her lifestyle to properly prevent the disease from progressing and the symptoms from recurring. Enhanced External Counterpulsation (EECP) therapy with its different mode of action provides a new treatment modality in the management of CAD (Fig. 10.1).

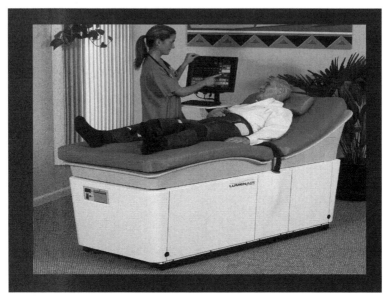

Figure 10.1 EECP Therapy: Consisting of a patient bed attached to an air compressor unit; computerized control consul; 3 sets of cuffs wrapped around the lower legs and the buttocks of the patient. (With permission from Vasomedical Inc., NY).
EECP: Enhanced External Counterpulsation

WHAT IS EECP?

EECP therapy is a noninvasive outpatient therapy consisting of electrocardiography (ECG)-gated sequential leg compression, which produces hemodynamic effects similar to those of an intra-aortic balloon pump (Fig. 10.2).

However, unlike intra-aortic balloon pump therapy, EECP therapy also increases venous return (Manchanda and Soran 2007, Birtwell et al. 1968). Cuffs resembling oversized blood pressure cuffs—on the calves and lower and upper thighs, including the buttocks—inflate rapidly and sequentially via computer interpreted ECG signals, starting from the calves and proceeding upward to the buttocks during the resting phase of each heartbeat (diastole). This creates a strong retrograde counterpulse in the arterial system, forcing freshly oxygenated blood toward the heart and coronary arteries while increasing the volume of venous blood return to the heart under increased pressure. Just before the next heartbeat, before systole, all three cuffs deflate simultaneously, significantly reducing

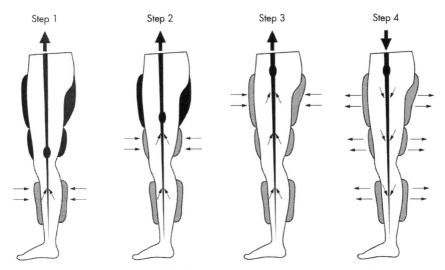

Figure 10.2 Technique of EECP Therapy.
Three pairs of pneumatic cuffs are applied to the calves, lower thighs, and upper thighs. The cuffs are inflated sequentially during diastole, distal to proximal. The compression of the lower extremity vascular bed increases diastolic pressure and flow and increases venous return. The pressure is then released at the onset of systole. Inflation and deflation are timed according to the R wave on the patient's cardiac monitor. The pressures applied and the inflation–deflation timing can be altered by using the pressure waveforms and ECG on the EECP therapy monitor. (With permission from Vasomedical Inc, NY).
ECG: Electrocardiogram;
EECP: Enhanced External Counterpulsation.

the heart's workload. This is achieved because the vascular beds in the lower extremities are relatively empty when the cuffs are deflated; significantly lowering the resistance to blood ejected by the heart and reducing the amount of work the heart must do to pump oxygenated blood to the rest of the body. A finger plethysmogram is used throughout the treatment to monitor diastolic and systolic pressure waveforms. A typical therapy course consists of 35 treatments administered for 1 hr a day over 7 wk.

EECP Therapy in CAD Management

Several nonrandomized and randomized trials have demonstrated a consistently positive clinical response among patients with CAD treated with EECP (Lawson et al. 1992, Arora et al. 1999). Benefits associated with EECP therapy include reduction in angina and nitrate use, increased exercise tolerance, favorable psychosocial effects, and enhanced quality of life as well as prolongation of the time to exercise-induced ST-segment depression and an accompanying resolution of myocardial perfusion defects (Masuda et al. 2001, Stys et al. 2002, Pettersson et al. 2006).

Most studies on EECP therapy cannot be double blind and lack good control groups because of technical limitations, drawbacks that have frequently raised questions regarding operator bias. However, the Multicenter Study of EECP (MUSTEECP), a randomized double-blinded placebo-controlled trial, did document a clinical benefit from EECP therapy in patients with chronic stable angina and positive exercise stress tests. Moreover, a MUST-EECP sub study demonstrated a significant improvement in quality-of life parameters in patients assigned to active treatment, and this improvement was sustained during a 12-mon follow-up (Arora et al. 2002). Although randomized (including placebo-controlled) and nonrandomized studies have shown beneficial effects of EECP therapy, investigators saw the need to assess EECP's effectiveness in real-world settings, leading them to develop the International EECP Patient Registry (IEPR) under the management of the University of Pittsburgh (Michaels et al. 2004). More than 5000 patients were enrolled in phase I and another 2500 patients enrolled in phase II of the study, and more than 90 centers participated. Results from the IEPR and the EECP Clinical Consortium have demonstrated that the symptomatic benefit observed in controlled studies translates to the heterogeneous patient population seen in clinical practice. Moreover, follow-up data indicate that the clinical benefit may be maintained for up to 5 yr in patients with a favorable initial clinical response (Lawson et al. 2000, Loh et al. 2008).

EECP Therapy in CAD with Left Ventricular Dysfunction

When providing EECP therapy to patients with heart failure, the initial researchers were concerned primarily that increased venous return resulting from EECP therapy might precipitate pulmonary edema in angina patients with severe left ventricular dysfunction (SLVD).

Using outcomes data from 363 patients enrolled in the IEPR, Soran et al. (Soran et al. 2006, Soran et al. 2002) evaluated the safety and efficacy of EECP therapy in those with refractory angina pectoris and SLVD (ejection fraction [EF] < 35%). They concluded that EECP therapy for angina is safe and effective in patients with SLVD who are not considered good candidates for revascularization by CABG) or PCI. After the patients completed therapy, there was a significant reduction in angina severity: 72% improved from severe angina to no or mild angina. Fifty-two percent of the patients stopped using nitroglycerin. There was also a significant increase in quality of life. At 2-yr follow-up, angina reduction was maintained in 55% of patients, the survival rate was 83%, and event-free (death/myocardial infarction [MI]/PCI/CABG) survival was 70%. Forty-three percent had no cardiac hospitalization; 81% had no congestive heart failure event (Fig. 10.3).

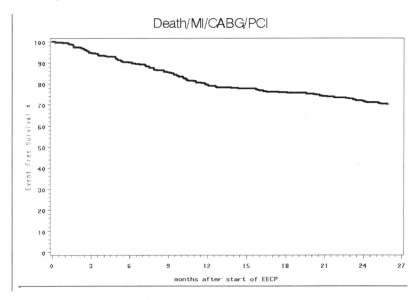

Figure 10.3 Event Free Survival Rate at 2 yr–Post EECP
Event (CABG; PCI; MI) free survival rate at 2 yr post EECP therapy in patients with coronary artery disease and SLVD (With permission from Dr. Soran; Soran et al. 2006).
CABG: Coronary Artery Bypass Grafting Surgery
MI : Myocardial Infarction
PCI: Percutaneaus Coronary Intervention
SLVD: Severe left ventricular dysfunction

Lawson et al. evaluated CAD patients with preserved left ventricular function (PLV; EF > 35%) and with SLVD (EF≤ 35%) who were treated with a 35-hr course of EECP. Bioimpedance measurements of cardiovascular function were obtained before the first and after the 35th hr of EECP therapy. Twenty-five patients were enrolled, 20 with PLV and five with SLVD. Angina class improved similarly in both groups. The SLVD group, in contrast to the PLV group, had increased cardiac power (i.e., mean arterial pressure × cardiac output/451), stroke volume, and cardiac index and decreased systemic vascular resistance with treatment. This study suggests that EECP may benefit patients experiencing CAD with SLVD directly by improving cardiac power and indirectly by decreasing systemic vascular resistance (Lawson et al. 2002).

Patients with CAD and left ventricular dysfunction exert an enormous burden on health care resources, primarily because of the number of recurrent emergency department visits and hospitalizations. Improvements in symptoms and laboratory assessments in these patients may not correlate with a reduction in emergency room visits and hospitalizations. Soran et al. assessed the impact of EECP therapy on emergency department visits and hospitalization rates at 6-mon follow up. In this prospective cohort study, clinical outcomes, number of all-cause emergency department visits, and hospitalizations within the 6 mon before EECP therapy were compared with those at 6-mon follow-up. The mean number of emergency department visits per patient decreased from 0.9 ± 2.0 before EECP to 0.2 ± 0.7 at 6 mon ($P < 0.001$), and hospitalizations were reduced from 1.1 ± 1.7 to 0.3 ± 0.7 ($P < 0.001$) (Soran et al. 2007).

MECHANISM OF ACTION

Upon diastole, cuffs inflate sequentially from the calves, raising diastolic aortic pressure proximally and increasing coronary perfusion pressure. Compression of the vascular beds of the legs also increases venous return. Instantaneous decompression of all cuffs at the onset of systole significantly unloads the left ventricle, thereby lowering vascular impedance and decreasing ventricular workload. This latter effect, when coupled with augmented venous return, raises cardiac output.

In summary, EECP therapy increases venous return, raises cardiac preload, increases cardiac output, and decreases systemic vascular resistance (Michaels et al. 2002). Mode-of-action studies have shown that EECP therapy increases angiogenesis factors such as human growth, basic fibroblast growth, and vascular endothelial growth factors. Enhanced diastolic flow increases shear stress, increased shear stress activates the release of growth factors, and augmentation of growth factor release activates angiogenesis (Masuda et al. 2001).

To test the hypothesis that EECP augments collateral function Gloekler et al randomized patients with chronic stable angina to EECP therapy or sham treatment. Before and after 30 hr of randomly allocated EECP or sham EECP, the invasive collateral flow index (CFI) was obtained in 34 vessels. CFI was determined by the ratio of mean distal coronary occlusive pressure to mean aortic pressure with central venous pressure subtracted from both. Additionally, coronary collateral conductance (occlusive myocardial blood flow per aorto-coronary pressure drop) was determined by myocardial contrast echocardiography and brachial artery flow-mediated dilatation was obtained. CFI significantly improved in the EECP group but not in the sham treatment. EECP appeared to be effective in promoting coronary collateral growth. The extent of collateral function improvement was related to the amount of improvement in the systemic endothelial function (Gloekler et al. 2010).

EECP therapy improves endothelial function and enhances vascular reactivity. Braith et al investigated the effects of EECP on peripheral artery flow-mediated dilation in a randomized placebo-controlled study. Symptomatic patients with CAD were randomized (2:1 ratio) to 35 1-hr sessions of either EECP or sham EECP. Flow-mediated dilation of the brachial and femoral arteries was performed with the use of ultrasound. Plasma levels of nitrate and nitrite, 6-ketoprostaglandin, F1α, endothelin-1, asymmetrical dimethylarginine, tumor necrosis factor-α, monocyte chemoattractant protein-1, soluble vascular cell adhesion molecule, high-sensitivity C-reactive protein, and 8-isoprostane were measured. EECP significantly increased brachial and femoral artery flow-mediated dilation, the nitric oxide turnover/production markers nitrate and nitrite and 6-keto-prostaglandin F1α, whereas it decreased endothelin-1 and the nitric oxide synthase inhibitor asymmetrical dimethylarginine in treatment versus sham groups, respectively.

EECP significantly decreased the proinflammatory cytokines tumor necrosis factor-α, monocyte chemoattractant protein-1, soluble vascular cell adhesion molecule-1, high-sensitivity C-reactive protein, and the lipid peroxidation marker 8-isoprostane in treatment versus sham groups, respectively. EECP also significantly reduced angina classification in treatment versus sham groups, respectively. Findings from this study provide novel mechanistic evidence that EECP has a beneficial effect on peripheral artery flow-mediated dilation and endothelial-derived vasoactive agents in patients with symptomatic CAD (Braith et al. 2010). As with athletic training, the vascular effects of EECP therapy may be mediated through changes in the neurohormonal milieu. Wu et al. (Wu et al. 1999) showed that EECP therapy has a sustained, dose-related effect in stimulating endothelial cell production of the vasodilator nitric oxide (NO) and in decreasing production of the vasoconstrictor endothelin. In another study, Qian et al. (Qian et al. 1999) showed that the NO level increased linearly in proportion to the hours of EECP treatment. Urano et al. (Urano et al. 2001) further showed that plasma brain natriuretic peptide levels decreased after EECP therapy and were positively correlated with left ventricular end diastolic pressure and negatively correlated with peak filling rate. They concluded that EECP therapy reduces exercise-induced myocardial ischemia in association with improved left ventricular diastolic filling in patients with CAD. Another possible mechanism explaining EECP's mode of action is that it may affect changes in ventricular function independent of changes in cardiac load. Gorcsan et al. (Gorcsan et al. 2000) evaluated the effects of EECP therapy on left ventricular function in New York Heart Association class II or III heart failure patients with an EF less than 40%. Their results showed that EECP treatment was associated with improvements in preload adjusted maximal power, a relatively load-independent measure of left ventricular performance and EF, along with a decrease in heart rate in these heart failure patients.

A recently published randomized controlled study examining the effect of EECP therapy on inflammatory and adhesion molecules in patients with CAD indicated that EECP therapy has an anti-inflammatory effect in patients with angina pectoris. Patients were randomly assigned to receive active EECP or sham treatment. Plasma tumor necrosis factor- α, monocyte chemoattractant protein-1, and soluble vascular cell adhesion molecule-1 were measured before and after a full course of 35 1-hour sessions of EECP or sham treatment. Patients in the EECP group demonstrated reductions in tumor necrosis factor-α and monocyte chemoattractant protein-1 after treatment, whereas those in the sham therapy group showed no changes. EECP therapy decreased circulating levels of proinflammatory biomarkers in patients with symptomatic CAD (Casey et al. 2008) (Fig. 10.4).

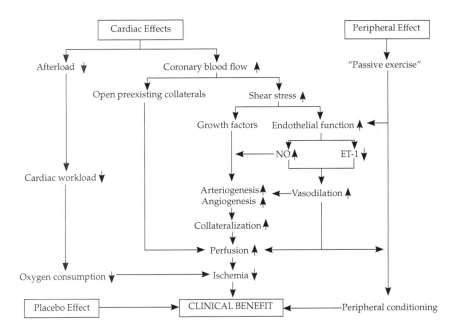

Figure 10.4 Mechanism of EECP Therapy.
Possible mechanisms responsible for the clinical benefit associated with enhanced external counterpulsation (EECP) therapy. Acute afterload reduction decreases myocardial demand. By increasing coronary blood flow, EECP therapy is thought to promote myocardial collateralization via opening of preformed collaterals, arteriogenesis, and angiogenesis. Increased blood flow and shear stress may also improve coronary endothelial function favoring vasodilation and myocardial perfusion. In addition, improvement in endothelial function may further promote collateral formation by arteriogenesis and angiogenesis. Besides a peripheral training effect, a minor placebo effect is considered to contribute to the symptomatic benefit of EECP therapy. (With permission from Dr. Bonetti; Bonetti et al. 2003).
ET =endothelin;
NO =nitric oxide.

PATIENT SELECTION

Indications

The following patients are considered good candidates for EECP (Braverman 2009).

1. Canadian Cardiovascular Society Class II–Class IV angina who are not readily amenable to surgical intervention such as PCI or CABG.
2. Patients who underwent incomplete surgical intervention for symptomatic CAD.
3. Patients at high risk of adverse events from invasive revascularization such as individuals with complex coronary anatomies, pulmonary disease, renal failure, advanced heart failure, diabetes mellitus, and those who are elderly and frail.
4. Patients with cardiac syndrome X or microvascular angina.
5. Patients with symptomatic ischemic cardiomyopathy.
6. Patients who restrict activities and have functional limitations to avoid angina.
7. Patients who are symptomatic despite maximal medical therapy or have undesirable side effects from their anti-anginal medications.

CONTRAINDICATIONS AND SIDE EFFECTS

Table 10.1 summarizes the side effects and contraindications of EECP Therapy (Table 10.1) (Manchanda and Soran 2007).

PRECAUTIONS

Few conditions preclude a patient from receiving EECP. Many of these are temporary or may be addressed to allow for EECP at a later date (Braverman 2009, Braverman 2005). Patients with sinus tachycardia or atrial fibrillation with a heart rate above 100 beats per minute should be rate-controlled prior to EECP. Such a rapid heart rate makes it difficult to achieve adequate diastolic augmentation as the time during diastole is markedly shortened. In addition, the quick pace of cuff inflation and deflation is uncomfortable for the patient.

Table 10.1 Side effects and Contraindications of EECP Therapy.

Side Effects
• Leg or waist pain • Skin abrasion or ecchymoses • Bruises in patients using coumadin when INR dosage is not adjusted • Paresthesias • Worsening of Heart Failure in patients with severe arrhythmias
Contraindications • Coagulopathy with INR of prothrombin time >2.5. • Arrhythmias that may interfere with triggering of EECP system (uncontrolled atrial fibrillation, flutter, and very frequent premature ventricular contractions) • Within two weeks after cardiac catheterization or arterial puncture (risk of bleeding at femoral puncture site) • Decompensated heart failure. • Moderate to severe aortic insufficiency (regurgitation would prevent diastolic augmentation) • Severe peripheral arterial disease (reduced vascular volume and muscle mass may prevent effective counterpulsation, increased risk of thromboembolism) • **Contraindications-continue**
• Severe hypertension >180/110 mm Hg (the augmented diastolic pressure may exceed safe limits) • Aortic aneurysm (≥ 5cm) or dissection (diastolic pressure augmentation may be deleterious) • Pregnancy or women of childbearing age (effects of EECP therapy on fetus have not been studied) • Venous disease (phlebitis, varicose veins, stasis ulcers, prior or current deep vein thrombosis or pulmonary embolism) • Severe chronic obstructive pulmonary disease (no safety data in pulmonary hypertension)

Summary of the side effects and contraindications of EECP Therapy. (With permission Dr Soran; Manchanda and Soran 2007).

INR: international normalized ratio

EECP: Enhanced External Counterpulsation

Patients at high risk of complications from increased venous return should be carefully chosen and monitored during EECP. In particular, patients with impaired left ventricular function may be thrown into an episode of acute pulmonary edema if the increased venous return during EECP is not off-set by proper afterload reduction.

Severe aortic insufficiency, aortic stenosis, or mitral stenosis may not allow for the appropriate hemodynamic changes that occur during EECP. Patients with these conditions should not undergo

EECP as increasing diastolic augmentation may worsen their clinical situation. However, after the valve is surgically repaired or replaced, a patient may receive EECP safely.

Pacemakers and internal defibrillators do not interfere with the efficacy of EECP. Patients with a rate-adaptive pacemaker may need the rate-responsive feature to be disengaged throughout treatment to avoid activation of a paced tachycardia from the body motion during EECP.

A fever is an indication that one may have bacteremia or viremia. The increased blood flow that results from exercise and from EECP could promote seeding of the infection in the heart. Just as someone should not exercise if they have a fever, they should not receive EECP until their temperature returns to normal.

If a patient has an open wound on one of the lower extremities where the EECP cuffs are applied, treatment should be delayed until the skin heals. However, non-healing ulcers on the feet do not interfere with EECP. The treatment improves blood flow in the legs, which helps to stimulate wound-healing. EECP should be delayed in superficial phlebitis until the inflammation resolves to ensure patient comfort.

PRACTICE AND PROCEDURES

1. EECP Medical Record

The patient's medical record should include a signed consent for EECP and a physician's prescription. A current physician evaluation including medical history, physical examination, ECG, and other pertinent tests as necessary should be included. All relevant previous cardiac medical records, including cardiac catheterizations, invasive procedures, stress tests, nuclear studies, echocardiogram, vascular studies, and current laboratory data should be included. A medication list of all current prescription and over-the-counter medications should be incorporated.

2. Specific Patient Instructions for EECP Treatment

Prior to beginning EECP, each patient should receive instructions that foster understanding of the treatment and maximize patient comfort.

Patients should take all their medications on their usual schedule. Delaying medications could result in the patient forgetting to take them which may lead to various adverse outcomes.

Individuals should not eat or drink any substantial amounts 90 min before treatment to maximize abdominal comfort and minimize the number of bathroom breaks. They may have a light snack such as crackers, a piece of fruit, etc. if desired.

Patients should wear spandex or Lycra tight-fitting pants with the inseam stitching on the outside to minimize skin irritation. Skin irritation is the most common adverse event related to EECP and prevention is a high priority. They should wear a loose-fitting shirt to allow for easy ECG electrode application and form-fitting underwear to reduce the risk of wrinkled fabric. Nylon pantyhose should be worn under EECP pants if the patient is at risk of skin breakdown (i.e., dry skin, peripheral vascular disease, diabetes, etc.). Moisturizing lotion should be applied to the legs several times daily to reduce skin dryness.

The patient should urinate immediately prior to EECP to minimize interruptions during the treatment hour. During EECP, renal blood flow increases augmenting urine production so breaks to empty one's bladder are common.

When patients are relaxed, arterial dilation and blood flow are improved. Music, television, or reading may serve as an excellent distraction.

3. Treatment Cuffs

EECP cuff placement and wrapping is the cornerstone of successful EECP therapy. The goal is to maximize the volume of compression while minimizing patient discomfort and skin irritation. Cuffs should be wrapped tightly without wrinkles.

There is a systematic approach to wrapping. First, the therapist must place the cuffs open on the treatment bed and position the patient supine so that his/her buttocks are centered in the upper cuff, the top of the cuff waistband falls at the patient's waist, and the abdominal strap is secured. Next the EECP pants are stretched toward the ankle and over the feet to ensure no wrinkles are under the cuffs. The upper thigh cuffs are wrapped high into the inguinal region to ensure proper compression of the femoral blood vessels. Finally, the calf cuffs are positioned under the calves, one to two fingerbreadths below the lower border of the patella, ensuring the lower edge of the cuff is proximal to the ankle malleoli. A 25 centimeter piece of foam padding is placed over the shin and the cuff straps are tightly wrapped. Strategically placed foam padding and/or sheep skin may be conservatively utilized to maximize patient comfort and protect sensitive areas of skin, stomas, abdominal catheters, etc. that may be aggravated by cuff compression.

4. ECG Electrodes

Meticulous skin preparation and correct lead placement with high-quality electrodes are vital to an artifact-free ECG signal and a comfortable treatment. Initially cleanse the skin of anticipated lead placement with alcohol and/or a lightly abrasive skin prepping gel. Connect each ECG wire securely to an electrode. The goal is to obtain an upright R wave with the maximum amplitude and the least amount of motion artifact. To achieve this, it is best to apply leads over bony prominences and leave slack to the ECG wires to prevent tension. The R wave is used as a reference point for triggering EECP. Inflation occurs at the onset of diastole, on or near the T wave while deflation occurs prior to the onset of systole, on or near the P wave. There is an inflation delay of 150 millisec after triggering the R wave and automatic deflation 30 millisec before the next anticipated R wave to prevent cuff inflation during systole. If R waves are not identifiable due to artifact or an arrhythmia, the system stops and stays in deflation mode. When three clear consecutive R waves are identifiable, counterpulsation resumes.

The standard electrode placement is:
- Ground lead at the right 4th or 5th rib, mid-clavicular to axillary line.
- Negative lead on the right clavicle or acromion.
- Positive lead at the left 4th or 5th rib, mid-clavicular to axillary line.

Lead placement should be adjusted to get the clearest rhythm strip with the highest amplitude. If possible, lead placement should be varied to avoid skin irritation. If there is excessive electrical artifact on the rhythm strip, the areas with hair that interfere with good skin contact should be shaved, the skin cleaned again and electodes reapplied firmly. Ensure the patient avoids touching the chest or moving their arms as these motions may create interference. If none of these steps rectify the problem, it may be helpful to try different lead placement. Examples include sternal placement (three leads on sternum with negative most cranial, ground in center, and positive most caudal), acromial placement (negative and positive leads on acromial prominences and ground on sternum), and anterior/posterior placement (negative lead on sternum, positive lead in sub-scapular region on either side of the back, ground on an acromion).

Finger Plethysmograph

The finger plethysmograph records the arterial tracing from the finger, which reflects the blood flow in all of the arteries in the body, including the heart. To get this reading accurately, the finger plethysmograph clip must be applied with the infrared sensors touching the ventral part of finger tip. The patient should relax the hand and hold the arm still. The arterial waveform amplitude and timing are fine-tuned as needed (discussed below). The EECP therapist should perform periodic checks, usually 20 min and 40 min into the treatment, and adjust the timing as needed.

If the patient is not able to achieve sufficient diastolic augmentation, it is essential to investigate the cause and remediate it. Ensure the proper cuff size is in use, the cuffs are tightly wrapped, and the inflation and deflation markers are adjusted for optimal diastolic augmentation. The cuff and system hoses must be checked for leaks or kinks and the bladders tested for leaks. Verify that the patient's hand and arm are relaxed and positioned properly for the finger plethysmograph reading. The amplitude of the arterial waveform may be increased as needed. If the problem persists, warm the hand and clean the fingertip before re-applying the finger plethysmograph clip and try moving it to another finger or to the other hand to get a better reading.

Augmentation is difficult to get with a rapid heart rate (> 90 beats per minute). Common causes of sinus tachycardia in an EECP patient include pain, anxiety, or the need to urinate, and oftentimes when these underlying causes are addressed appropriately, the tachycardia resolves.

EECP Waveforms

The finger plethysmogram tracing is used to set, monitor and adjust timing of EECP therapy and to quantify the hemodynamic effects of counterpulsation. During EECP, cuff inflation and deflation change the arterial waveform so that the diastolic peak is elevated, indicating diastolic augmentation, while the end diastolic pressure and the systolic peak are lowered, demonstrating systolic unloading. Measurement of augmentation is based upon the ratio of diastolic (D) to systolic (S) wave, or the D/S ratio. The D/S ratio may be measured in terms of area or peak (Fig. 10.5).

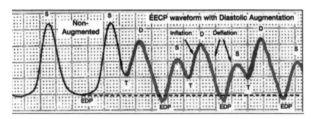

Figure 10.5 The Finger Plethysmogram Tracing.
The finger plethysmogram tracing: infrared photo-electric sensors used to record changes in pulsatile blood flow from the finger. (Unpublished data).
S: systolic wave
D: diastolic wave
EDP: end diastolic pressure
T: beginning of inflation

Color image of this figure appears in the color plate section at the end of the book.

The peak measurement (P) is more common as it is easily done by estimation. The peak of the S is given the value of 1.0 and the peak of the D is compared to this. If D is taller than S, then $P > 1.0$, if the D is equal to S, then $P = 1.0$, and if D is shorter than S, then $P < 1.0$. The goal of EECP is $P \geq 1.0$, which is called therapeutic diastolic augmentation. Buttons on the EECP therapy system allow for inflation and deflation timing changes to maximize diastolic augmentation.

The initial timing is set by placing the inflation timing marker on the peak of the T wave and the deflation timing marker on the peak of the p wave on the ECG. Further timing adjustments may then be made.

The goal of cuff inflation timing is to increase diastolic amplitude, indicating an increase in coronary perfusion and venous return. Early inflation may impinge upon systole, decrease cardiac output and increase cardiac workload. Late inflation shortens the interval of cuff compression and counterpulsation, decreasing coronary perfusion and venous return. The goal of cuff deflation timing is to make systolic unloading easier and decrease cardiac workload. Early deflation leads to early decompression of the lower extremities and shortened duration of counterpulsation. Late deflation does not allow time for the vasculature of the legs to refill. They are still compressed as systole begins, increasing peripheral resistance and cardiac workload.

If $P < 1.0$, the causes that must be fixed include movement of the patient's finger or body during measurement, dirty or cold fingers, low applied treatment pressure, loose cuff wrap, or poor timing of inflation and/or deflation. Some physiological factors cause $P < 1.0$ such as peripheral vascular disease, diabetes, congestive heart failure, anemia, low blood pressure, and smoking. In these patients, higher augmentation may not be possible. Such individuals often require more than the standard 35 EECP treatments to achieve their maximum clinical benefit.

Other treatment guidelines

Patients should be monitored continuously by trained medical personnel during EECP.

KEY FACTS OF CAD AND EECP THERAPY

- CAD is a narrowing or blockage of the arteries and vessels that provide oxygen and nutrients to the heart. Chest pain (angina pectoris) is the most common symptom of CAD.
- Almost 2500 Americans die of cardiovascular disease each day, an average of one death every 35 sec. Cardiovascular disease claims more lives each year than the next four leading causes of death combined—cancer, chronic lower respiratory diseases, accidents, and diabetes mellitus.
- There are three commonly used approaches for treating CAD: Medication; PCI and CABG.
- EECP therapy with its different mode of action provides a new treatment modality in the management of CAD.
- Although the concept of counterpulsation was introduced in the United States in the early 1950s, it took more than 40 yr for investigators to develop the effective technology that is currently being used.
- EECP therapy is a non-invasive outpatient treatment for patients with chronic angina pectoris
- Increases myocardial perfusion by the application of pressure to the lower legs, thighs and buttocks to relieve the workload on the heart, and encourages the growth of collateral circulation.
- Sequential three stage, computer controlled, operation
- A typical therapy course consists of 35 treatments administered for 1 hr a day over 7 wk.
- Benefits associated with EECP therapy include reduction in angina and nitrate use, increased exercise tolerance, favorable psychosocial effects, and enhanced quality of life as well as prolongation of the time to exercise-induced ST-segment depression and an accompanying resolution of myocardial perfusion defects.
- Mechanism of Action of EECP is through collateral development (development of new branches of blood vessels); endothelial (the inner lining of our blood vessels) function and neurohormonal improvement.

- Numerous clinical trials in the past two decades have shown EECP therapy to be safe and effective for patients with CAD, with a clinical response rate averaging 70 to 80%, which is sustained up to 5 yr.
- Need to undergo for a repeat EECP therapy is 13% at one year.

SUMMARY POINTS

- CAD is a narrowing or blockage of the arteries and vessels that provide oxygen and nutrients to the heart. Medication, PCI and CABG are the three most common treatment regimes used in CAD. Despite panoply of recent therapeutic advances, patients with CAD are not adequately treated; therefore, scientists have been investigating new technologies to help these patients.
- EECP therapy is a noninvasive outpatient therapy with a different mechanism of action than PCI and CABG, consisting of ECG-gated sequential leg compression. Cuffs resembling oversized blood pressure cuffs—on the calves and lower and upper thighs, including the buttocks—inflate rapidly and sequentially via computer interpreted ECG signals, starting from the calves and proceeding upward to the buttocks during the resting phase of each heartbeat (diastole). This forces oxygenated blood toward the heart and coronary arteries. Just before the next heartbeat, before systole, all three cuffs deflate simultaneously, significantly reducing the heart's workload.
- Benefits associated with EECP therapy include reduction in angina and nitrate use, increased exercise tolerance, favorable psychosocial effects, and enhanced quality of life as well as prolongation of the time to exercise-induced ST-segment depression and an accompanying resolution of myocardial perfusion defects. It shows its long term effectiveness through collateral artery development, endothelial function and neurohormonal improvement. Seventy to 80% of patients respond the treatment and 66% maintain the improvement over 5 yr.

- Indications: Patients who underwent incomplete surgical intervention for symptomatic CAD; patients at high risk of adverse events from invasive revascularization such as individuals with complex coronary anatomies, pulmonary disease, renal failure, advanced heart failure, diabetes mellitus, and those who are elderly and frail; patients with cardiac syndrome-X or microvascular angina; patients with symptomatic ischemic cardiomyopathy; patients who restrict activities and have marked functional limitations in order to avoid anginal symptoms; patients who are symptomatic despite maximal medical therapy or have undesirable side effects from their anti-anginal medications.
- Contraindications: bleeding diathesis (for patients on Coumadin, goal INR ≤ 3.0); active deep venous thrombophlebitis (assess with lower extremity Doppler ultrasound); severe lower extremity vaso-occlusive disease (assess with ankle-brachial index test and lower extremity Doppler ultrasound); presence of a documented aortic aneurysm requiring surgical repair (assess with an abdominal ultrasound); pregnancy (assess with venous blood test for beta-hCG level)
- Major side effects of the EECP therapy are: leg or waist pain; skin abrasion or ecchymoses; bruises in patients using Coumadin when INR dosage is not adjusted; paresthesias; worsening of heart failure in patients with severe arrhythmia

DEFINITIONS

Angina/Angina Pectoris: symptom indicating the heart is not receiving enough blood and oxygen. Angina may be chest pain, pressure, and/or chest tightness. Other 'anginal equivalent' symptoms include shortness of breath, fatigue, decreased exercise tolerance, burning or discomfort in the chest or throat, pain in the jaw, neck, arms, upper back, or peri-scapular region.

Atherosclerosis: the process whereby deposits of fat, cholesterol, and calcium build up in the arteries as plaques, leading to the blockages in coronary artery disease and other cardiovascular conditions.

Cardiovascular Disease: a family of diseases of the heart and circulatory system, including hypertension, coronary artery disease, stroke, and congestive heart failure.
Collateral Blood Vessels: small vessels that develop over time in response to increased blood flowing through them; often a result of narrowed coronary arteries.
Coronary Artery Disease (CAD): The narrowing of the arteries that supply blood to the heart usually caused by atherosclerosis.
Diastole: the resting phase of the cardiac cycle, during which the heart receives 80% of its blood supply.
Diastolic Blood Pressure: the lowest blood pressure measured in the arteries during diastole.
Enhanced External Counterpulsation (EECP): a non-invasive treatment for heart disease whereby pneumatic cuffs are wrapped around the patient's legs and inflate at the onset of diastole, then deflate just prior to systole. As a result, the treatment increases myocardial blood flow, improves cardiac function, decreases the myocardial workload, and decreases systemic vascular resistance.
Ischemia: deficiency of blood and oxygen, signaling that oxygen demand is greater than supply.
Refractory Angina: persistent, severe angina in patients who are inoperable and on maximum medications.
Stable Angina: angina that occurs at predictable times and levels of exertion and usually lasts 20 to 30 min or less. It may continue without significant change for years. Rest and/or nitroglycerin provide short-term relief.
Systemic Vascular Resistance: the resistance blood encounters as it flows out of the left ventricle and throughout the body.
Systole: the contraction phase of the cardiac cycle, during which the heart sends blood throughout the body.
Systolic Blood Pressure: the highest pressure measured in the arteries during systole.
Sydrome-X: presence of angina without clear atherosclerosis on coronary angiography.
ST segment: the part of an ECG immediately following the QRS complex and merging into the T wave.

LIST OF ABBREVIATIONS

Beta-hCG	:	beta subunit of human chorionic gonadotropin
CABG	:	coronary artery bypass surgery
CAD	:	coronary artery disease
CFI	:	coronary flow index
EECP	:	enhanced external counterpulsation
ECG	:	electrocardiogram
EF	:	ejection fraction
ET	:	endothelin
Fig.	:	figure
INR	:	international normalized ratio
MI	:	myocardial infarction
No	:	nitric oxide
PCI	:	percutaneous coronary interventions (stent implantation or balloon angioplasty)
PLV	:	preserved left ventricular function
SLVD	:	severe left ventricular dysfunction

REFERENCES CITED

Arora, R.R., T.M. Chou, D. Jain, B. Fleishman, L. Crawford, T. McKiernan and R.W. Nesto. 1999. The Multicenter Study of Enhanced External Counterpulsation (MUST-EECP): Effect of EECP on exercise-induced myocardial ischemia and anginal episodes. J Am Coll Cardiol 33: 1833–1840.

Arora, R.R., T.M. Chou, D. Jain, B. Fleishman, L. Crawford, T. McKiernan, R. Nesto, C.E. Ferrans and S. Keller. 2002. Effects of enhanced external counterpulsation on health-related quality of life continue 12 months after treatment: a substudy of the multicenter study of enhanced external counterpulsation. J Investig Med 50: 25–32.

Birtwell, W.C., U. Ruiz, H.S. Soroff, D. DesMarais, and R.A. Deterling. 1968. Technical consideration in the design of a clinical system for external left ventricular assist. Trans Am Soc Artif Intern Organs 14: 304–310.

Braith, R.W., C.R. Conti, W.W. Nichols, C.Y. Choi, M.A. Khuddus, D.T. Beck and DP Casey. 2010. Enhanced external counterpulsation improves peripheral artery flow-mediated dilation in patients with chronic angina. A randomized sham-controlled study. Circulation 122: 1612–1620.

Braverman, D.L. 2005. EECP is a modern marvel—there is no cutting on the cutting edge. pp.14–32 *In:* DL Braverman (ed). Heal your Heart with EECP: the only Way to Overcome Heart Disease. Celestial Arts, Berkeley, California, USA.

Braverman, D.L. 2009. Enhanced external counterpulsation: An innovative physical therapy for refractory angina. PM & R 1: 268–76.

Casey, D.P., C.R. Conti, W.W. Nichols, C.Y. Choi, M.A. Khuddus and R.W. Braith. 2008. Effect of enhanced external counterpulsation on inflammatory cytokines and adhesion molecules in patients with angina pectoris and angiographic coronary artery disease. Am J Cardiol 101: 300–302.

Gloekler, S., P. Meier, S. F de Marchi, T. Rutz, T. Traupe, S.F. Rimoldi, K. Wustmann, H. Steck, S. Cook, R. Vogel, M. Togni and C. Seiler. 2010. Coronary Collateral Growth by External Counterpulsation: A Randomized Controlled Trial Heart 96: 202–207.

Gorcsan, J., L. Crawford and O.Z. Soran. 2000. Improvement in left ventricular performance by enhanced external counterpulsation in patients with heart failure. J Am Coll Cardiol 35: 230A.

Lawson, W.E., J.C.K. Hui, H.S. Soroff, S.Z. Zeng, D.S. Kayden, D. Sasvary, H. Atkins and P.F. Cohn. 1992. Efficacy of enhanced external counterpulsation in the treatment of angina pectoris. Am J Cardiol 70: 859–862.

Lawson, W.E., J.C.K. Hui and P.F. Cohn. 2000. Long-term prognosis of patients with angina treated with enhanced external counterpulsation: five-year follow-up study. Clin Cardiol 23: 254–258.

Lawson, W.E., K. Pandey and J.C.K. Hui. 2002. Benefit of enhanced external counterpulsation in coronary patients with left ventricular dysfunction: cardiac or peripheral effect? J Card Fail 8(Suppl 41): 146.

Loh, P.H., J.G. Cleland, A.A. Louis, E.D. Kennard, J.F. Cook, J.L. Caplin, G.W. Barsness, W.E. Lawson, O.Z. Soran and A.D. Michaels. 2008. Enhanced external counterpulsation in the treatment of chronic refractory angina: a long-term follow-up outcome from the International Enhanced External Counterpulsation Patient Registry. Clin Cardiol 31:159–164.

Lloyd-Jones, D., R.J. Adams, T.M. Brown, et al. 2010. Heart Disease and Stroke Statistics 2010 Update: A Report from the American Heart Association. Circulation 121: e466–e215.

Manchanda, A. and O. Soran. 2007. Enhanced external counterpulsation and future directions: step beyond medical management for patients with angina and heart failure. J Am Coll Cardiol 50: 1523–1531.

Masuda, D., R. Nohara and K. Kataoka. 2001. Enhanced external counterpulsation promotes angiogenesis factors in patients with chronic stable angina. Circulation 104: II445.

Masuda, D., R. Nohara, T. Hirai, K. Kataoka, L.G. Chen, R. Hosokawa, M. Inubushi, E. Tadamura, M. Fujita and S. Sasayama. 2001. Enhanced external counterpulsation improved myocardial perfusion and coronary flow reserve in patients with chronic stable angina. Eur Heart J 22: 1451–1458.

Michaels, A.D., M. Accad, T.A. Ports and W. Grossman. 2002. Left ventricular systolic unloading and augmentation of intracoronary pressure and Doppler flow during enhanced external counterpulsation. Circulation 106: 1237–1242.

Michaels, A.D., G. Linnemeier, O. Soran, S.F. Kelsey and E.D. Kennard. 2004. Two-year outcomes after enhanced external counterpulsation for stable angina pectoris (from the International EECP Patient Registry [IEPR]). Am J Cardiol 93: 461–464.

Pettersson, T., S. Bondesson, D. Cojocaru, O. Ohlsson, A. Wackenfors and L. Edvinsson. 2006. One year follow-up of patients with refractory angina pectoris treated with enhanced external counterpulsation. BMC Cardiovasc Disord 6: 28.

Soran, O., E.D. Kennard, S.F. Kelsey, R. Holubkov, J. Strobeck and A.M. Feldman. 2002. Enhanced external counterpulsation as treatment for chronic angina in patients with left ventricular dysfunction: a report from the International EECP Patient Registry (IEPR). Congest Heart Fail 8: 297–302.

Soran, O., E.D. Kennard, A.G. Kfoury and S.F. Kelsey. IEPR Investigators. 2006. Two-year clinical outcomes after enhanced external counterpulsation(EECP) therapy in patients with refractory angina pectoris and left ventricular dysfunction (report from The International EECP Patient Registry). Am J Cardiol 97: 17–20.

Soran, O., E.D. Kennard, B.A. Bart and S.F. Kelsey. IEPR Investigators. 2007. Impact of external counterpulsation treatment on emergency department visits and hospitalizations in refractory angina patients with left ventricular dysfunction. Congest Heart Fail 13: 36–40.

Stys, T.P., W.E. Lawson, J.C.K. Hui, B. Fleishman, K. Manzo, J.E. Strobeck, J. Tartaglia, S. Ramasamy R. Suwita, Z.S. Zeng, H. Liang and D. Wener. 2002. Effects of enhanced external counterpulsation on stress radionuclide coronary perfusion and exercise capacity in chronic stable angina pectoris. Am J Cardiol 89: 822–824.

Qian, X., W. Wu and Z.S. Zheng. 1999. Effect of enhanced external counterpulsation on nitric oxide production in coronary disease. J Heart Disease 1: 193

Urano, H., H. Ikedah, T. Ueno, T. Matsumoto, T. Murohara and T. Imaizumi. 2001. Enhanced external counter pulsation improves exercise tolerance, reduces exercise-induced myocardial ischemia and improves left ventricular diastolic filling in patients with coronary artery disease. J Am Coll Cardiol 37: 93–99.

Wu, G.F., S.Z. Qiang and Z.S. Zheng. 1999. A neurohormonal mechanism for the effectiveness of the enhanced external counterpulsation. Circulation 100: I832.

Index

3-hydroxy-3-methylglutaryl coenzyme A reductase inhibitors (statins) 48, 55

A

Acidosis 161
Acute coronary syndrome (ACS) 40, 43, 48
Acute myocardial infarction (AMI) 40
AHA/ACC guidelines 75
Angina 79, 85, 91
Angina Frequency 11, 17
Angina pectoris 1–3, 5, 17–19, 21, 22, 34, 35, 179, 182, 192, 194
angina stability scale 64
Angiotensin converting enzyme 91
anti-anginal drug 116, 122
Anti-ischemic 161
Antiplatelet therapy 87, 89, 91
Appropriateness 107
Aspirin 77, 79, 82, 85, 87, 88
ATP 155, 158–162, 169–171

B

beta blockers (BBs) 82, 88, 89, 114, 116–121, 123, 127–130

C

CADENCE 7, 8, 10, 12, 19
calcium channel blockers (CCBs) 114, 116, 119, 120, 129–130
Cardiac arrest 77
Cardiac computed tomography 95–98, 101, 109
Chest pain 21–35, 77, 91
Chronic Stable Angina 1, 3, 5, 12, 16–18, 115, 116, 126
Collateral development 192
Coronary anomalies 103, 108
Coronary artery bypass 78, 91, 92
Coronary artery bypass graft (CABG) 78–80, 85, 87, 92, 108
Coronary Artery Disease 5, 7, 8, 10, 19, 174, 175, 179, 194–196
Coronary plaque rupture 39, 40, 54
Cost effectiveness 106, 110

C-reactive protein (CRP) 46, 49, 50, 54
Cross-cultural issues 66

D

Defibrillation 77
Definite angina 62
Direct thrombin inhibitors 133, 134, 136, 139, 140, 145, 150–152
disease-specific questionnaires 58, 60

E

EECP 174–196
Effort angina 163
Electrocardiogram (ECG) 77–81, 83, 88, 90, 92
Endothelial function 175, 181, 183, 193
Enhanced external counterpulsation 174–177, 183, 185, 195, 196
Extracardiac findings 105

F

Fatty acid oxidation 121, 123, 155, 156, 158, 162, 167–170
Fibrinolysis 82, 84, 86, 91

G

Gastro-intestinal 22, 26, 28, 35
generic questionnaires 60
Glucose oxidation 121–123, 155, 156, 158, 160, 161, 167–170

H

Health status 7, 9, 10
Health-related quality of life 7
Heart failure 178, 179, 182, 184, 185, 191, 194, 195
hemodynamic agents 114, 116, 126, 128, 129
Heparin 85, 86, 92
Hybrid imaging 107, 110

I

Imaging 96, 100, 101, 102, 106–110
Incidence 1, 5–7, 9, 13, 18

Indication 109
Intravascular Ultrasound (IVUS) 45, 48, 54, 102
Ischemia 85, 91
ivabradine 121–123, 128, 129, 131

L

Lactate 158, 171
Left ventricular dysfunction 178–180, 196
Left ventricular function 103, 104, 110
Lifestyle 32
London School of Hygiene Cardiovascular Questionnaire 61

M

Matrix metalloproteinases (MMPs) 43, 52, 55
Medical Outcomes Study 36-Item Short Form Health Survey 58, 60, 65, 67, 68
medical treatment 116
Metabolic agent 121, 168
Molecular imaging 107, 108
Multidetector computed tomography (MDCT) 48, 49, 54
Multi-slice computed tomography 95, 96, 102, 104, 109
Musculoskeletal 25, 26, 28, 35
Myocardial infarction 2–4, 12, 13, 18, 74, 75, 79, 80, 86, 89–92
Myocardial Ischaemia 1, 3, 4, 9, 17–19

N

Neurohormonals 182, 192, 193, 198
nicorandil 121, 123, 124, 128
nitrates 116, 117, 119, 121, 124, 127–129
Non-cardiac chest pain 21, 25–28, 30–35
Non-invasive 95, 96, 99, 103, 106, 109
Nottingham Health Profile questionnaire 58, 60, 65, 67, 70

O

Optical Coherence Tomography (OCT) 45–48, 53– 55
optimal medical therapy (OMT) 116, 129, 130
Oxidized-low density lipoprotein (Ox-LDL) 17, 55
oxygen supply 115, 121, 126

P

Percutaneous coronary intervention (PCI) 78, 79, 83, 91, 92

Plaque 101, 102
Possible angina 63
Prevalence 1, 5–8, 13, 16, 18, 19
Psychological 27, 30, 31, 33
Pyruvate dehydrogenase 158, 159, 162

Q

quality of life 58–60, 64, 67–70
quality of life improvement 125

R

Radiation 96, 99, 103, 105, 106, 109
Randle cycle 121, 158, 160
ranolazine 121–124, 128
Refractory angina 179, 195
Reperfusion 80–83, 85, 87
Revascularization 114–116, 126, 179, 184, 194
Risk factors 21, 22, 24, 25, 30, 32, 34, 35
Rose Angina Questionnaire 58, 60–62, 67, 68, 70

S

Seattle Angina Questionnaire 11, 58, 60, 63, 67, 68, 701
Self-report 59
SF-36 58, 60, 65, 67, 68, 70
Spatial and temporal resolution 95, 96, 109
Stable angina pectoris (SAP) 43, 55, 58–60, 70
ST-Elevation myocardial infarction (STEMI) 74–90, 79, 90, 92
Stent 100, 101, 108, 109

T

Temporal resolution 95–97, 99, 109
Thin-cap fibroatheroma (TCFA) 42, 43, 45–48, 53, 55
Thrombin 133–142, 145, 147, 149–152
Trimetazidine 121, 128, 155–157, 160–168, 170

U

Unstable angina pectoris (UAP) 43, 47, 48, 55

V

Valve disease 104, 105
Vulnerable plaque 42, 48–50, 53

About the Editors

Colin R. Martin BSc PhD RN YCAP CPsychol CSci AFBPsS is a qualified Nurse and Chair in Mental Health at the University of the West of Scotland and Adjunct Professor at the Royal Melbourne Institute of Technology (RMIT), Melbourne, Australia. He is also a Chartered Health Psychologist and a Chartered Scientist and has worked in senior management posts in the NHS followed by academic posts in the UK and the Far East. He has conducted original research in both the addictions and the mental health aspects of chronic diseases. Professor Martin is honorary Consultant Psychologist to The Salvation Army, UK and Eire Territories and was instrumental in formulating the addictions policy of the Salvation Army (UK and Eire) over recent years to develop high quality and evidence-based clinical care and services. He has published many scientific papers in psychology, biology, medical and nursing journals. He is also Editor of several books.

Victor R. Preedy PhD, DSc, FSB, FRCPath, FRSPH is Professor of Nutritional Biochemistry, King's College London, Professor of Clinical Biochemistry, King's College Hospital (Honorary) and Director of the Genomics Centre, King's College London. Presently he is a member of the King's College London School of Medicine. In his career Professor Preedy has carried out research at the National Heart Hospital (part of Imperial College London) and the MRC Centre at Northwick Park Hospital. He has collaborated with research groups in Finland, Japan, Australia, USA and Germany. He is a leading expert on the mechanisms of disease and has lectured nationally and internationally. He has published in many peer-reviewed journals and has edited over 20 books.

Color Plate Section

Chapter 3

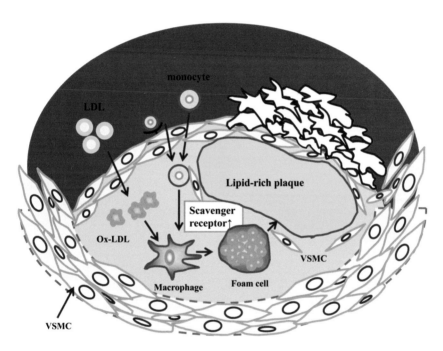

Figure 3.1 The evolution of the atherosclerotic plaque. Blood monocytes enter the arterial wall in response to chemoattractant cytokines such as monocyte chemoattracnt protein 1. Scavenger receptors mediate the uptake of modified lipoprotein particles and promote the development of foam cells. Atherosclerosis progresses through lipid core expansion and macrophage accumulation at the edges of the plaque, leading to fibrous cap rupture.

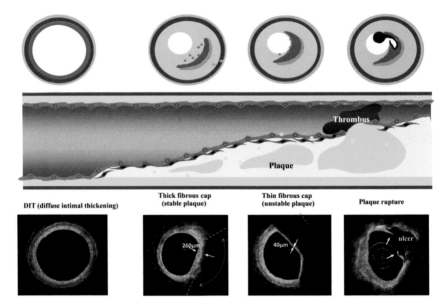

Figure 3.3. A generalized and progressive atherothrombosis process and corresponding optical coherence tomographic images. (Upper) This 'time-line' depiction of the development of atherothrombosis in a coronary vessel emphasizes the long-term nature of this process. (lower) Optical coherence tomography (OCT) provides high-resolution images of the coronary artery up to 10 μm.

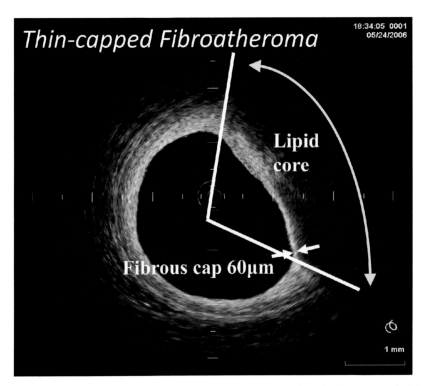

Figure 3.4. OCT-derived TCFA. A. A fibrous cap was identified as a signal-rich homogenous region overlying a lipid core, which was characterized by a signal-poor region on the OCT image. The lesion with a fibrous cap of <65μm was diagnozed as OCT-derived TCFA.

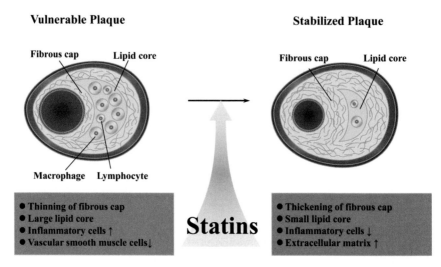

Figure 3.6. Effects of statins on plaque vulnerability. The pathological features of the most common form of vulnerable plaque include a large lipid pool within the plaque, a thin fibrous cap, and macrophage accumulation within the cap. Statins affect coronary plaque vulnerability, that is, the changes from vulnerable to stable plaque, which increase fibrous cap thickness and a decrease in total atheroma volume. (Adapted from Libby P, et al: Nat Med. 2002; 8: 1257–62.)

Chapter 6

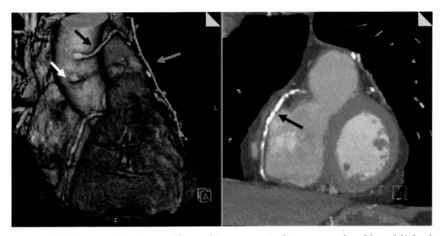

Figure 6.2 Bypass imaging with cardiac computed tomography. Unpublished images.

The left internal mammarian artery is connected to left anterior descending artery (grey arrow). The black arrow indicates a patent venous bypass graft to the circumflex artery whereas another venous bypass graft is occluded (white arrow). The heavily calcified right coronary artery is displayed on the right side (black arrow).

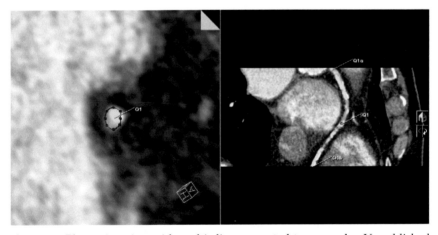

Figure 6.3 Plaque imaging with multi-slice computed tomography. Unpublished images.

Semiautomatic contour detection with post-processing software allows quantifying coronary plaque burden within the total vessel. On the left side cross-section of a coronary lesion ('Q1') which is located in the middle of the right coronary artery (right side). Different colours indicated different contrast attenuation of the plaque components.

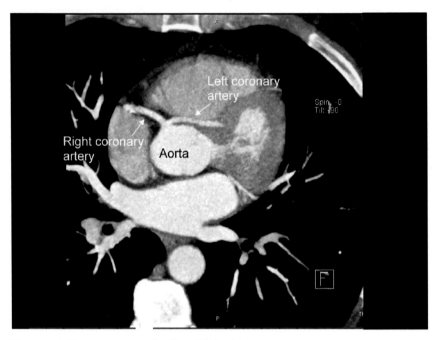

Figure 6.4 Coronary anomaly. Unpublished image.

This figure demonstrates a common ostium of the right and left coronary artery.

Chapter 7

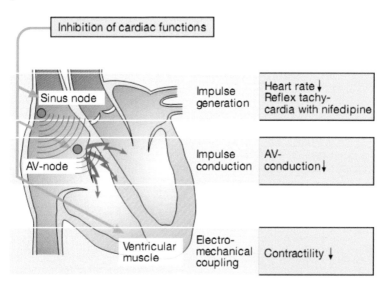

Figure 7.1 CCBS cause a reduction in heart rate (except for dihydropiridines which may cause reflex tachycardia), atrio-ventricular conduction velocity and cardiac inotropy. Abbreviations: CCBs (calcium channel blockers), AV (atrio-ventricular).

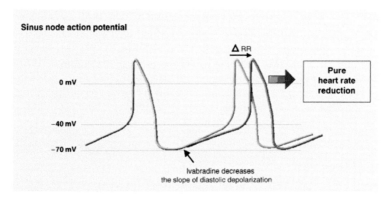

Figure 7.2 Ivabradine slows spontaneous activity of the senoatrial node (SAN) by blocking I_f-channels, thereby causing a decreased rate of diastolic depolarization (ΔR). This effect is not associated by a decrease in cardiac inotropy and therefore constitutes a pure heart rate decrease.

Chapter 8

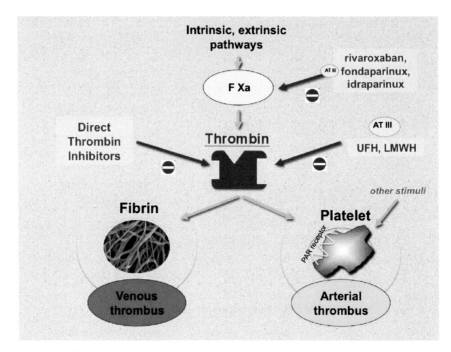

Figure 8.1 Thrombin central role for the formation of clots, with its antagonists.

Legend: the anticoagulant activity of unfractionated heparin (UFH) and low molecular weight heparins (LMWH) are mediated by the interaction with antithrombin III. Heparins generate a ternary heparin–thrombin–antithrombin complex, leading to thrombin inhibition. The activity of DTIs is independent of the presence of antithrombin and is related to the direct interaction of these drugs with the thrombin molecule. Unpublished.

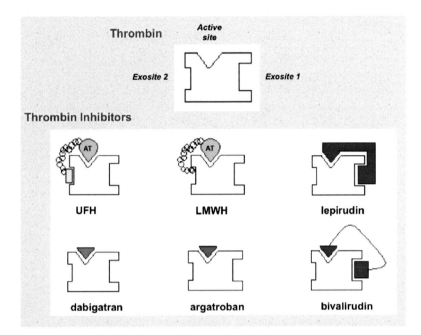

Figure 8.2 Thrombin molecule and interaction with mediators.

Legend: there are three domains on the thrombin molecule: an active site and two opposite located exosites. UFH binds exosite 2 and antithrombin III, which binds the enzyme active site. LMWH lack the longer chains of UFH that bind exosite 2. Lepirudin binds directly both the active enzymatic site and exosite 1 of thrombin, forming a slowly reversible complex. Bivalirudin binds thrombin in a fast, reversible fashion. Argatroban and dabigatran are small molecules that bind reversibly to the active enzymatic site of thrombin. Exosite 1=fibrin/fibrinogen binding site; Exosite 2=heparin-binding site; UFH=unfractionated heparin; LMWHs=low-molecular-weight-heparins; AT=antithrombin; DTI=direct trhombin inhibitor. Unpublished.

Chapter 10

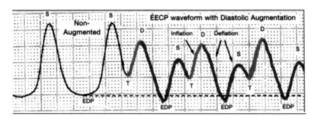

Figure 10.5 The Finger Plethysmogram Tracing.
The finger plethysmogram tracing: infrared photo-electric sensors used to record changes in pulsatile blood flow from the finger. (Unpublished data).

S: systolic wave

D: diastolic wave

EDP: end diastolic pressure

T: beginning of inflation